Electronics — A First Course

Electronics — A First Course

Owen Bishop

AMSTERDAM • BOSTON • HEIDELBERG • LONDON • NEW YORK • OXFORD
PARIS • SAN DIEGO • SAN FRANCISCO • SINGAPORE • SYDNEY • TOKYO
Architectural Press is an imprint of Elsevier

Newnes
An imprint of Elsevier
Linacre House, Jordan Hill, Oxford OX2 8DP
30 Corporate Drive, Burlington, MA 01803

First published 2002
Reprinted 2003, 2004

British Library Cataloguing in Publication Data
A catalogue record for this book is available from the British Library

Library of Congress Cataloguing in Publication Data
A catalogue record for this book is available from the Library of Congress

ISBN 0 7506 5545 3

For information on all Newnes publications
visit our website at www.newnespress.com

Printed and bound in Great Britain by MPG Books Ltd, Bodmin, Cornwall

Part 2: Electronic components

Design time   157

Design time   161

162

Design time   180

Supplements   182

Introduction

This is a complete introductory textbook intended for courses leading to GCSE Electronics. It also caters for the Electronics sections of GCSE Design and Technology (Electronic Products), for the Electricity and Magnetism sections of GCSE Physics, the GNVQ unit on Electronics for Engineering, and the Intermediate GNVQ unit on Electronics Principles and Applications. It assumes no previous knowledge of electronics or of the electronic aspects of physics.

Most of the text is divided into numerous short Topics, presented as double page spreads. These cover the topics common to the majority of the specifications at the level of the Foundation Tier (FT). The topics are dealt with thoroughly, with simple explanations and plenty of examples and illustrations. This double-page presentation has two advantages. It allows students to confine their attention to the particular topics found in a given specification. It also presents the student with self-contained, easily assimilable and readily testable segments of knowledge.

At frequent intervals throughout the text there are sets of Extension Boxes. These contain the material that is found only in the Higher Tier, and also deal with topics that appear in only one or two of the specifications. Students may read or omit these boxes according to the specification and tier for which they are working.

The book is student-centred and it features:

- Frequent 'Self Test' questions to allow students to assess their progress.

- Sets of questions at numerous key stages in the book, linking together the material of consecutive Topics. Answers to these and the Self Test questions are given at the end of the book, if they are numerical or of few words.

- An abundance of practical examples, with numerous circuit diagrams. Detailed instructions for simple constructional techniques, and for test procedures appear in the 'In the Lab' sections.

- An underlying emphasis on electronics systems.

- 'Design time' pages provide a wealth of practicable suggestions for circuits and projects that students can design for themselves. These are intended to help students to prepare coursework projects for the examination, as well as to promote understanding of electronic theory.

- The book is copiously illustrated with half-tone and line drawings, the circuit symbols following the guide set out in the AQA GCSE Electronics specification. There are numerous photographs, including close-ups of electronics components, and illustrations of constructional techniques.

What is electronics?

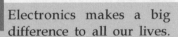

Electronics is about electrons.

It is about the ways we use electrons to do useful and interesting things.

The photos on this page show some of the things that we can do with electrons. Can you work out what these things are?

Electronics makes a big difference to all our lives. Without electronics, our lives would be less comfortable, less safe, less interesting and less fun. Or do you disagree? Talk about it with other students.

How to use this book

Most of the book is set out as *double-page spreads*. When you open the book at one of these spreads, the two pages tell you all you need to know about one electronic topic. As well as the text and pictures, there may be:

- **Things to do** that help you to learn more.
- **Design tips** to help you design and build your electronic project.
- **Self tests** to find out how you are getting on. Answers are at the back of the book.

Most students will need to work on most of the spreads. These cover all the essential topics for Electronics exams at Level 2. Your teacher will tell you which spreads, if any, you can leave out.

In the Lab and **Design Time** pages provide you with advice and ideas for your project.

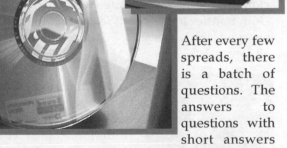

After every few spreads, there is a batch of questions. The answers to questions with short answers are at the end of the book.

Also after each group of spreads are some pages that give you more details or cover extra topics. These are mostly for students taking the Higher Tier papers in the exam. They also cover topics that are required by only one or two of the exam boards. Ask your teacher which topics to study.

1 Electrons

Put some small pieces of kitchen foil on the workbench. You can use small pieces of cork, instead. Rub a plastic pen with a dry woollen cloth. Rub hard for ten or twenty seconds. Hold the pen a few millimetres above the pieces of foil. They jump up and stick to the pen. Some of them may jump up and down again several times.

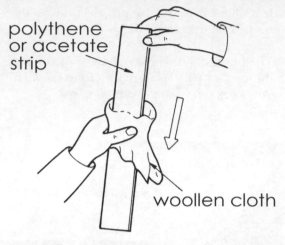

polythene or acetate strip

woollen cloth

The reason that the pieces jump is that they are attracted by electrons on the pen. Rubbing the cloth on the pen has made electrons from the cloth transfer to the pen. We say that the pen is **charged** with electrons. It has an **electric charge**.

Some other things can be charged by rubbing. Rub a balloon with a cloth (or against your clothes). Then place it in contact with the wall of the room. It does not fall down to the floor but stays where you put it, on the wall. The electric charge has produced an **electric force** that holds the balloon against the wall.

Things to do

You need two strips of polythene, about 30 cm by 2 cm, and a soft dry cloth. Put the strips on the workbench and rub them briskly with the cloth. Pick up the strips by one end, one in each hand. Hold them about 50 cm apart. Then slowly move them together.

Repeat this, using one strip of polythene and one strip of acetate sheet. What do the strips do now?

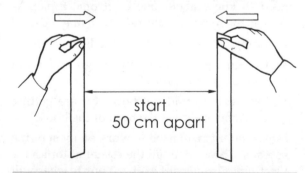

start
50 cm apart

Self test

What do you expect will happen if you try to bring two charged acetate strips together?

Kinds of charge

You have found that:

- Two charged polythene strips **repel** each other. They try to stay apart.
- A polythene strip and an acetate strip **attract** each other. They try to come together.

It seems that the charge on acetate is different from that on polythene, so:

There are two kinds of charge

Two charged polythene strips repel each other, so:

Like charges repel

Two differently charged strips attract each other, so:

Unlike charges attract

Self test

Pieces of foil jump up to a charged plastic pen. Then some of them jump down again. Why does this happen?

Positive and negative charge

The two kinds of charge are called **positive charge** and **negative charge.** These names do not mean that positive charge has something that negative charge does not have. They just mean that the charges are of opposite kinds.

Rubbing a polythene strip with a cloth transfers some of the electrons from the atoms in the cloth on to the strip. Electrons have negative charge, so the strip becomes negatively charged. Also, the atoms of the cloth have now lost some electrons. This makes the cloth positively charged.

Rubbing an acetate strip with a cloth does the opposite. It *removes* electrons from the strip, leaving it positively charged. The cloth gains electrons and becomes negatively charged.

Using energy

Positive and negative charges always attract each other. They try to come together. When you rub the cloth on the plastic, you separate the negative charge from the positive. It takes energy to pull them apart when they are trying to come together. This energy comes from the muscles of your arm.

Electrons

Electrons are too small to see, even with a powerful microscope.

Electrons are too light to weigh. You need 1 000 000 000 000 000 000 000 000 000 000 electrons to weigh 1 kg (an amazing fact that you do not need to remember).

The most important fact about electrons is that they carry **negative electric charge**. The charge on a single electron is extremely small. But, if you have enough of them (as on the pen or the charged polythene), you can show the force that their charge causes. There are lots more things that we can do with electrons, as you will find out as you work through the book.

Self test

Why is it impossible to have a *pile* of electrons, like that shown in the drawing?

3

2 Static and current electricity

When two different substances rub together, electrons are transferred from one substance to the other. One substance gains electrons and becomes negatively charged. The other loses electrons and becomes positively charged.

However, this happens only with certain substances, such as polythene, acetate, wool, and rubber. Other substances that can be charged in this way include glass, ceramics, nylon, paper, and air. These substances are known as **insulators**. The electrons stay fixed on the surface of the substance and can not move away. Insulators do not let charge flow through them. We say the charge is **static,** a word meaning 'standing'.

A person may become charged when walking on a carpeted floor. The rubbing of their plastic soles against the carpet (often nylon) generates a charge on the person. In very dry climates the charge on the person may become very large. They feel a 'tingle', hear a tiny 'tick' or may even see a spark, when they touch an earthed metal object, such as a door handle.

Similarly, just sitting working at a desk may generate charge because of the person's woollen, nylon or polyester clothes rubbing together. Electronics engineers have to be particularly careful to avoid this, as the charge they pick up on their bodies may destroy delicate electronic components when they touch them.

Lightning

The charges build up so much that there is a strong attraction between the electrons and the nearest positive charges. A lightning flash occurs as electrons rush across the gap. The air along the flash is heated so much that it causes a shock wave, which we hear as thunder.

Small particles in a cloud collide together. It is thought that the larger particles gain electrons and become negatively charged. The smaller particles lose electrons and become positively charged. Movements in the cloud tend to sort out the particles by size. The top of the cloud has a positive charge, and the bottom a negative charge.

Most lightning flashes are within a cloud, and some are between a cloud and one of its neighbours. They do not do any damage, except possibly to aeroplanes flying through the cloud.

The most dangerous flashes are between a cloud and charged areas on the ground.

Conductors

Some substances let electric charge flow through them. These substances are called **conductors**.

One of the best-known conductors is copper. It conducts so well because the electrons of copper atoms are able to escape easily from the atoms.

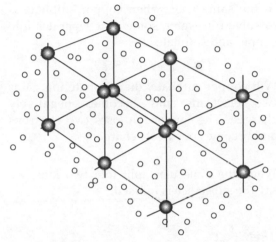

In a piece of copper, the atoms (large spheres in the drawing above) are arranged in regular rows and columns, called a **lattice**. The electrons that have escaped from the atoms (small circles) are able to wander about freely in the space between the copper atoms.

If we connect a battery to each end of a strip of copper, its negative terminal supplies electrons to the copper. Its positive terminal removes electrons from the other end. They are attracted by the positive (opposite) charge.

Things to do

Test different substances to find out if they are conductors or insulators (non-conductors).

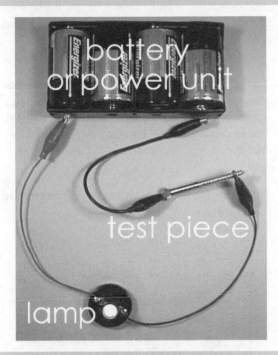

Try this with different materials, such as: an iron nail or screw (as shown), a piece of brick, a copper strip or wire, a plastic rod, a strip of aluminium kitchen foil, a piece of wood, a 'silver' coin, the 'lead' of a pencil (not really lead, but carbon), a piece of stone, and other materials..
The lamp shines when the material is a conductor. Make lists of conductors and non-conductors.

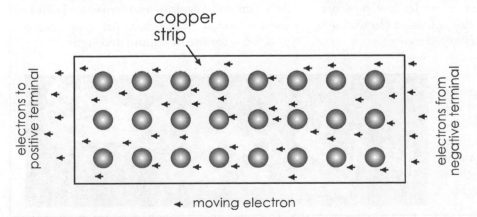

copper strip

electrons to positive terminal

electrons from negative terminal

◄ moving electron

The flow of electrons along the copper strip is called an **e l e c t r i c current**. The flow is from negative to positive.

Ions in solutions

When a salt such as sodium chloride is dissolved in water, some of its molecules separate into two atoms, sodium and chlorine:

$$NaCl \rightarrow Na + Cl$$

An electron from the sodium becomes attached to the chlorine. The chlorine (gaining an electron) is negatively charged. The sodium (losing an electron) is positively charged:

$$NaCl \rightarrow Na^+ + Cl^-$$

The charged sodium and chlorine atoms are called **ions**.

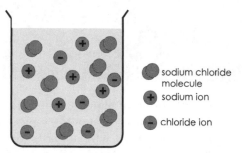

- sodium chloride molecule
- sodium ion
- chloride ion

Memo

Symbols for elements:
Cl = chlorine   O = oxygen
Cu = copper   S = sulphur
Na = sodium

In the same way, when copper sulphate is dissolved in water, some of it separates into copper and sulphate ions:

$$CuSO_4 \rightarrow Cu^{++} + SO_4^{--}$$

Wih copper sulphate, the sulphate ion gains two electrons and becomes negatively charged. The copper ion loses two electrons and becomes positively charged.

When the molecules split into two ions, we say the substance is **ionised.**

Self Test

Which of these represent ions?

Cl^-, $CuSO_4$, Na^+, H_2O, SO_4^{--}, H_2, $NaCl$.

Ions in gases

Gases can be ionised too. A neon striplight consists of a sealed glass tube containing neon gas. The neon is under low pressure. Some of the neon atoms lose an electron and become positively charged ions:

$$Ne \rightarrow Ne^+ + e^-$$

When a current is passed through the tube, the electrons move rapidly along the tube striking the neon ions. The ions glow with a bright red light.

Neon lamps are used for signs and also as low-power indicator lamps. Other gases, such as argon and krypton, also ionise and can be used in lamps. These gases produce a range of colours for use in illuminated signs.

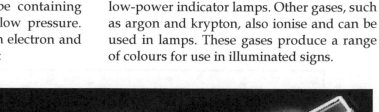

Electric current

An electric current is a flow of negative charge (electrons) from negative to positive.

In electronics, we usually think of the current as flowing from positive to negative. Although this is not what actually happens, most people like to think of it in that way. This idea of a current flowing from positive to negative is known as **conventional current**. In the rest of this book, when we say 'current' we mean conventional current, flowing from positive to negative.

Elecctrons are negative **charge carriers.**

When current flows through a gas (opposite) or through a solution (see next page) there may be positive charge carriers as well. These are the positive ions such as neon (Ne^+), sodium (Na^+) and copper (Cu^{++}).

The electrons carry negative charge from negative to positive. At the same time, the positive ions carry positive charge from positive to negative. There is a two-way conduction of charge.

Conductors

The best conductors are metals. Copper is the most commonly used conductor because it conducts electric charge better than any metal, except silver. But silver is too expensive to be used. Copper wires are used in almost all electronic equipment. The tracks on a circuit board are also made of copper.

The next best conductor is aluminium. This is often used in power lines, because of its lightness and cheapness. It is not as strong as copper, so a few strands of steel wire are included when making the cable.

Carbon is a non-metal but it has important uses as a conductor. It does not conduct charge as well as the metals do. Rods of carbon are used for making certain kinds of electric cell. Carbon is also used in making resistors (p. 30).

Solutions of salts in water are reasonably good conductors (see *Ions in solutions*, opposite). Much of the human body consists of such solutions, so the body is a reasonable conductor of electricity. This is why we must be very careful when handling electrical equipment and working with electricity in the laboratory. Even quite a small current through a part of the body can paralyse the nerves and may kill you. Electricity can also cause unpleasant burns.

Self test

List these substances in order, from the best conductor to the worst:

aluminium, rubber, copper, carbon, gold, silver.

Discharging static electricity

A charged plastic pen or balloon may hold its charge for an hour or more. The charge is on an insulating substance that stops it from flowing away. But the charge can be removed by contact with the air. The air often contains charged ions. A positively charged ion in the air is attracted toward an electron on a charged object. When it contacts it, the electron transfers to the ion and their charges cancel out. Gradually, all the electrons on the surface of the charged object are removed in this way. The object becomes discharged.

Charged conductors lose their charge quickly if they are connected to ground by a wire. Electronics engineers often wear metallised wrist bands, connected by wire to ground. This allows any charge on the body of the engineer to be conducted away to ground, and so prevents static charges from damaging the electronic components.

Sometimes a charge builds up on an object faster than it can leak away or can be conducted away. The charge may eventually become so great that it breaks though the air as a spark. Lightning is the most powerful example of this.

An aeroplane picks up static charge as it flies through the air. On landing, the plane does not remain charged, because its tyres are made from special conductive rubber. If it were not for these, there would be a danger of fire or explosion resulting from electrostatic sparks while refuelling.

Self test

In what three ways may a charged object lose its charge?

Conduction through solutions

A solution of substances that ionise is called an **electrolyte**. If we place two conducting rods (or **electrodes**) in an electrolyte and connect a battery or power unit to them, conventional current flows through the solution from positive to negative.

Copper ions (positive) are attracted toward the negative electrode (**cathode**). On arrival, they gain two electrons each and are discharged. The electrons have come from the negative terminal of the battery. The discharged ions are deposited on the cathode as metallic copper. This process is called **electrolysis**.

Sulphate ions (negative) are attracted toward the positive electrode (**anode**). There they are discharged, giving up two electrons, which flow to the battery.

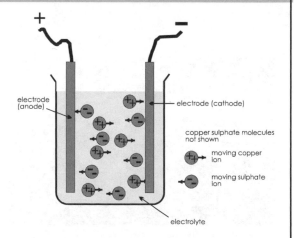

It can be shown by weighing that the amount of copper deposited on the cathode is proportional to the size of the current and to the length of time for which the current passes. This shows that copper ions carry a fixed amount of charge.

Photocopying Extension Box 7

To make a photocopy, place the original document face-down on the glass and close the cover. Press the START button. The photocopy is made in six stages:

1 A high voltage is applied to the **corona wire 1.** Positive ions from the corona wire are attracted toward the **drum**. The drum is a cylinder of aluminium, coated with a special light-sensitive layer.

2 A bright light shines on the original document and a system of lenses focuses an image of the document on to the drum.

3 The light in the brighter parts of the image causes electrons to appear on the light-sensitive layer. These *discharge* the positive ions already there. There is less effect in the darker parts of the image. There is now an image on the drum, in which the darkest parts still have positive ions. The lightest parts have none.

4 A black powder (toner) is scattered on to the drum. It is negatively charged, so it is *attracted* toward the positive ions. It forms an image on the drum, black where the original was black.

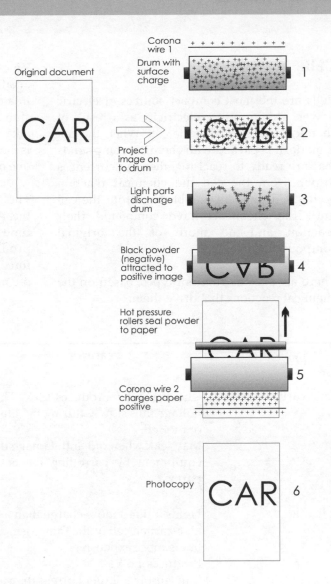

Original document

Corona wire 1
Drum with surface charge — 1

Project image on to drum — 2

Light parts discharge drum — 3

Black powder (negative) attracted to positive image — 4

Hot pressure rollers seal powder to paper

Corona wire 2 charges paper positive — 5

Photocopy — 6

5 A sheet of paper is fed past the drum as it turns. Before reaching the drum, the paper is positively charged by another corona wire. The charge on the passing paper strongly *attracts* the powder, which is still negatively charged. The attraction to the paper is stronger than the attraction to the drum, so the powder is transferred to the paper. Next the paper passes between two hot pressure rollers. These melt the powder which then sticks firmly to the paper. This produces an image on the paper which is an exact copy of the original image.

Many copiers have a lens system that projects an image larger or smaller than the original. Then the photocopy is identical to the original except for its size.

There is not enough room on the drum to copy a whole sheet at once. The drum turns several times to copy one sheet, a little at a time. The drum is automatically cleaned and discharged after the part-image has been transferred to the paper. Then it is re-charged (stage 1) ready to receive the next part of the image.

3 Cells and batteries

Cells

Cells are the most compact sources of electric power. They produce electricity as a result of chemical actions inside the cell. When the cell is made, it is packed with chemical compounds that are ready to react together. As current is drawn from the cell the chemical reaction occurs. Current can be obtained from the cell until the chemicals have completed their reaction and no more of the original compounds are left.

There are several different types, based on the chemical reactions that drive them:

Zinc-carbon cells and alkaline cells are made in a number of different standard sizes. The alkaline cell on the right is the AAA size, which is one of the smallest.

The photograph shows the AAA cell about 1.4 times its real size. It produces an electrical force of 1.5 volts (1.5 V, see next topic).

Type of cell	Features	Examples of uses
Zinc-carbon	Cheapest type. Produces 1.5 V. The voltage falls slowly during the life of the cell. May leak when old and damage the equipment by corroding its metal parts.	Electrric torches, handlamps, doorbells, security alarms
Alkaline	Holds 3 times more charge than a zinc-carbon cell of the same size, but is more expensive. Produces 1.5 V. Can supply a larger current than a zinc-carbon cell Voltage steady during life of cell; falls sharply at the end (keep a spare handy) Does not leak.	Electronic equipment such as clocks, remote controllers, electronic toys. Also used for the same purposes as zinc-carbon cells.
Silver oxide	Made as button cells. Produces 1.4 V. Delivers a small current for a long time.	Digital watches, small pocket calculators.
Lithium	Produces a small current for a long time (several years) but can produce a large current for a short time.	Memory backup in computers. Pocket calculators.

Rechargeable cells

For a continuous power supply, or occasional large bursts of current, we use rechargeable cells. When the cell is exhausted, we connect it to a charger, powered from the mains supply. It restores the chemicals in the cell to their original state. Rechargeable sources include:

Danger!

NiCad and lead-acid batteries can produce very large currents. They can give a dangerous electric shock. A 12 V lead-acid car battery is capable of providing far more current than a 12 V battery made up from zinc-carbon cells. Take care when working with NiCad and lead-acid batteries.

Type of cell	Features	Examples of uses
Nickel-cadmium (NiCad)	Store less charge than zinc-carbon cells of the same size. Produce 1.2 V. Voltage falls rapidly when nearly discharged. Can deliver high current.	High-current portable equipment such as video cameras, digital cameras, mobile phones.
Lead-acid (Accumulator)	Produce 2 V. Can deliver very high current. Not portable; lead electrodes make the cell heavy. Danger of spilling acid electrolyte.	Car starter motors.

Batteries

A battery is a number of cells connected together. They are usually connected so that the battery produces an increased output voltage.

For example, the popular PP3 battery produces 9 V. It is made up from six cells, each delivering 1.5 V. It is used in clocks, test meters, and in other low-current equipments.

A battery can be built up from separate cells in a **battery box**. The plastic box has contacts and wires and to join the cells together. The box in the photograph on page 5 has four alkaline cells. Each produces 1.5 V, so the total voltage of the battery is 6 V.

This compact battery of 8 alkaline cells is only 28 mm long. It produces a 12 V supply. Small batteries such as this are used where we need a high voltage in a small space. Examples include photographic equipment and key-fob remote controllers.

Self Test

1 A lead-acid car battery produces 6 V. How many cells does it contain?

2 A torch battery is made up of 3 alkaline cells. What voltage does it produce?

3 What type of cell or battery would you select for powering (a) a mini-computer game, (b) a hearing aid (c) a flying model aeroplane, and (d) a petrol lawn-mower motor?

4 Current, voltage and power

Current

The unit of electric current is the **ampere**. Its symbol is **A**. Few people use the word 'ampere'. They use '**amp**' instead. A typical mains electric lamp needs about one-third of an amp to make it glow brightly. A two-bar room heater needs about 8 amps.

Smaller amounts of current are measured in **milliamps**. A milliamp, symbol **mA,** is one thousandth of an amp. A torch bulb uses 60 mA or less.

An even smaller unit of current is the **microamp,** symbol **μA**. A microamp is one thousandth of a milliamp, or one millionth of an amp. Electronic clocks and watches take only a few microamps. A single AAA cell (p. 10) can power a digital clock for many months.

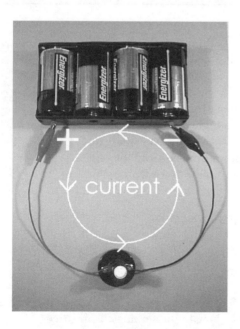

Current flows round a circuit. It flows out of the positive terminal of the battery, through the lamp and back into the battery through the negative terminal. The same amount of current flows in all parts of the circuit.

Voltage

Voltage is the *electric force* that drives current around an electric circuit. The unit of voltage is the **volt**, symbol **V**. Most cells produce a voltage of about 1.5 V. The voltage of the mains supply is 230 V.

At a power station, the voltage is higher and is measured in **kilovolts**, symbol **kV**. A kilovolt is equal to one thousand volts. On high-power transmission lines the voltage may be as high as 400 kV.

Small voltages are measured in **millivolts,** symbol **mV**. A millivolt is one-thousandth of a volt. Voltages that are smaller still are expressed in **microvolts,** symbol **μV**. A microvolt is one-thousandth of a millivolt or one-millionth of a volt. Electrical signals from a microphone and other sensors are generally measured in millivolts or microvolts.

Current is driven round the circuit by the electric force (the voltage) between the positive and negative terminals of the battery.

Power

Electric power expresses the rate at which an electrical device is converting energy from one form into another. For example a room heater converts electrical energy into heat energy. The rate at which it does this is its power, expressed in watts. The symbol for watt is **W**.

An average electric lamp runs at 75 W. A typical two-bar heater is rated at 2000 W, or 2 **kilowatts**. The power of an electrical power station (converting energy from coal into electrical energy) is measured in **megawatts**.

It can be shown that the power of a device is proportional to the amount of current flowing through it. It is also proportional to the voltage that is driving the current. The bigger the current and the bigger the driving force, the bigger the power. Writing this as an equation:

power = current × voltage

- power is expressed in watts
- current is in amps
- voltage is volts

Example 1

An electric steam iron runs on the 230 V mains and the current through it is 5.65 A. What is its power?

power = 230 × 5.65 = 1299.5 W

This result is close to 1300 W, or 1.3 kW.

Example 2

A 100 W lamp is running on the mains at 230 V. What current is passing through it?

Rearranging the power equation above gives:

current = power/voltage

Using this version of the equation:

current = 100/230 = 0.435 A

Current is 0.435 A, or 435 mA.

Self test

1 A projection lamp runs on a 36 V supply and takes 11 A. What is its power?

2 An electric hair-clipper runs on 230 V, and is rated at 10 W. How much current does it take?

3 A torch uses two alkaline cells and the current through the lamp is 60 mA. What is the power of the torch?

4 A 25 W soldering iron runs on 50 V. How much current does it use?

5 Express **(a)** 34 A in mA, **(b)** 1.2 mA in µA, **(c)** 1.2 mA in A, **(d)** 5505 mA in A, and **(e)** 58 µA in mA.

6 Express **(a)** 4.5 V in mV, **(b)** 11 kV in V, **(c)** 675 mV in V, **(d)** 521 µV in mV, **(e)** 0.55 V in mV, **(f)** 440 µV in V **(g)** 0.22 mV in V, and **(h)** 3300 V in kV.

7 Express **(a)** 675 W in kW, **(b)** 25 MW in kW, **(c)** 650 mW in W, **(d)** 6 MW in W, **(e)** 4450 kW in MW, **(f)** 2.55 W in mW, **(g)** 79 kW in W, and **(h)** 33 MW in kW.

Summing up

Electrical quantity	Units of measurement	Symbols
Current	**amp** (ampere)	**A**
	milliamp	mA
	microamp	µA
Voltage	kilovolt	kV
	volt	**V**
	millivolt	mV
	microvolt	µV
Power	megawatt	MW
	kilowatt	kW
	watt	**W**
	milliwatt	mW

The basic units are in heavy type. The multiple and submultiple units that are listed are the ones most often used in electronics.

13

In the Lab Using a multimeter

Measuring current and voltage

A **multimeter** measures several different electrical quantities, usually including voltage and current. There are two kinds of multimeter, analogue (left below) and digital (right).

Both meters have sockets for **probes**:

- Positive, marked '+' and usually red.
- Common (negative), marked COM or '–' and usually black.

Some meters have a second positive socket which **must** be used for high voltages.

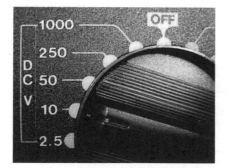

Both kinds of meter have a knob for selecting which quantity to measure. It often selects the range too. On the right, the knob is selecting the 0 V to 10 V range.

With an analogue meter, you must touch the black (common) probe to the more negative point and red to the more positive. If you touch them the wrong way round, the needle swings *below* zero. Remove the probes from the test points immediately or you may damage the meter. Digital meters often have **autopolarity.** These work with the probes touching either way and display a minus sign to the left of the reading, if necessary.

With an analogue meter, there is the problem that the reading is wrong if you do not look straight down at the scale. This is called a **parallax error**. To help you avoid this, the meter has an arc of mirror on the scale. You can see the needle reflected in this. When you take a reading, move your head from side to side, until you have lined up the reflection of the needle exactly behind the needle. This makes sure that you are looking straight down on the scale and the reading will be correct.

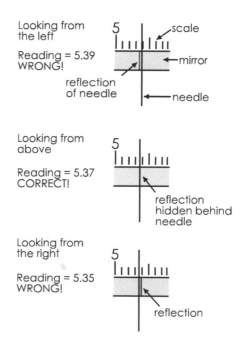

Digital meters (such as that in the photo) may be **autoranging**. They select the correct range automatically when a measurement is made.

Placing the probes

When you measure current, the meter is part of the circuit. The current flows through the meter. So, break the circuit at one point and connect the meter across the break.

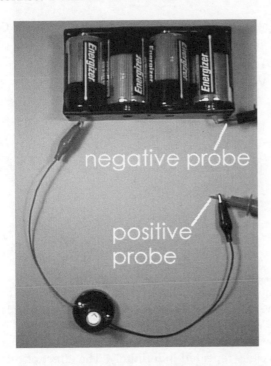

When you measure voltage, you are measuring the difference of electrical force at two points in the circuit. Do not break the circuit. Touch the probes to the two points to measure the difference.

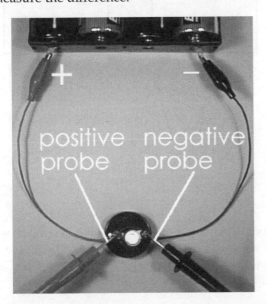

Test routine

1 *Before* you connect the meter to the circuit, set the selector knob to the range you want to use. If the meter has autoranging, you need only select the quantity (current or voltage). If you are not certain which range to choose, select the highest range (for example, 0 V to 1000 V). Then, if the meter shows only a very small reading, you can work down through the ranges to one that gives a readable result.

2 Connect the meter to the circuit, or just touch the probes to two points (remember about polarity – black to negative, red to positive).

3 Take the reading (remember to avoid parallax), and *write it down*.

4 If you need to change the range, disconnect the meter from the circuit first.

5 Disconnect the meter from the circuit when you have taken the reading.

6 Turn the selection knob to the OFF position.

Things to do

Set up circuits like those in the photos, but using cells or batteries of different voltages. You could also use some old batteries that have lost most of their charge. Use various lamps or, instead of the lamp, connect in a small electric motor, an electric bell or a buzzer.

For each circuit:

1 Make a simple drawing of the circuit.

2 Measure the current, and write the result on your drawing.

3 Measure the voltage across the lamp or other device, and write the result on your drawing.

Use your readings of current and voltage to calculate the power of the lamp or device.

5 Alternating currents

The voltage at the positive terminal of a battery stays constant until the the cell is exhausted. If we plot a graph of the voltage during a perod of time, the graph for a fresh battery is like this:

The graph is a horizontal straight line. It shows that the voltage of the battery is constant at 6 V.

If we connect the battery to a lamp, we can use a multimeter to measure the current flowing through it. Because the voltage is constant, the current that it drives is constant too. The graph of current against time is a straight horizontal line, like the graph above. Constant current like this is called **direct current**. This name is often shortened to **DC**.

DC always flows in the same direction, from positive to negative.

Memo

As usual, we are thinking of current as conventional current (p. 7).

The current that we get from some kinds of generator (including mains generators) is different from this. It repeatedly **changes its direction**.

This kind of current is called **alternating current**. This name is often shortened to **AC**.

Alternating voltage

The voltage at one terminal of a typical AC generator is shown in this graph:

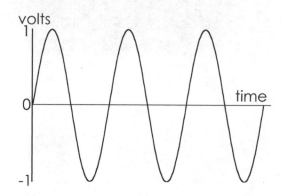

At one instant, the terminal is 1 V positive of the other. Then the voltage decreases until it is 1 V negative of the other terminal. Then it swings back again to become 1 V positive. This cycle is repeated indefinitely. The effect of this reversing voltage on the current in a circuit is shown below:

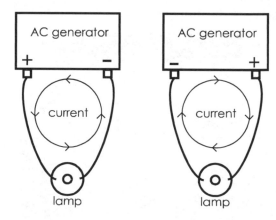

As the voltages at the terminals swing from positive to negative and back, the current repeatedly changes its direction. It is alternating current.

Describing an alternating current

A direct current is simple to describe. We just quote its size, in amps. To describe an alternating current we need to state three things:

- **Amplitude:** its maximum size, in amps.

- **Period:** the time for one complete cycle from positive back to positive again.

- **Waveform:** the shape of its graph.

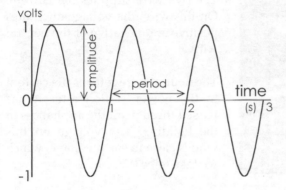

The AC plotted above has an amplitude $A = 1$ V. It has a period $P = 1$ s.

If we plot a graph of the sine of an angle from $0°$ to $360°$, it has exactly the same shape as one cycle of the AC graph. The AC has a sinewave shape. Not all AC has this shape, but this is the commonest. The mains AC and any AC from mains transformers are sinewaves.

Frequency

The unit of frequency is the **hertz**. Its symbol is **Hz**. The hertz is defined as a frequency of one cycle per second. The AC waveform shown in the graph has a frequency of 1 cycle per second, or 1 Hz.

If the period of a waveform is P seconds, there are $1/P$ cycles in one second. Its frequency is $1/P$ Hz:

$$\text{frequency} = 1/\text{period}$$

Self test

1 The frequency of the mains is 50 Hz. What is its period?

2 A musical note has a period of 3.9 ms. What is its frequency?

3 Describe this waveform:

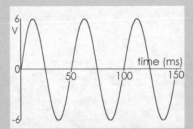

4 The 'bleep' of a microwave oven has a frequency of 1000 Hz. What is its period?

Things to do

1 Look at a demonstration of the shape of AC waveforms, using an oscilloscope. View the mains AC or the AC from mains transformers. Look at sinewaves of other frequencies and amplitudes from a signal generator.

2 Listen to a demonstration of the sound of sinewaves of various frequencies. This is best demonstrated by connecting an audio amplifier to the output of a signal generator. If possible, view and hear the signals at the same time.

3 Using the same equipment as for item 2 above, try to find out which is the lowest frequency that you can hear. Try to find out which is the highest frequency you can hear.

4 Using a microphone and audio amplifier connected to an oscilloscope, look at the shapes of sound waves produced by various musical instruments. Try to measure the frequency of notes on the musical scale.

6 Mains electricity

Mains electricity is produced in power stations like the one illustrated in the photo below.

This is one of the two generators at Ironbridge Power Station in Shropshire. Here the generators are driven by turbines. The turbines are turned by steam under pressure. The steam is produced in a coal-fired boiler. The rate of production of electrical energy from the chemical energy in coal is over 200 MW.

In other power stations, the steam may be produced in an oil-fired boiler, or by using heat from a nuclear reactor. In a hydro-electric power station, the turbines are turned directly by the flow of water. Wind is another source of energy for electricity generation.

In a few parts of the World, such as the thermal areas of New Zealand, steam for turbines is generated using energy from the hot rocks just below the Earth's surface.

In the United Kingdom and many other countries, the mains supply to the consumer is close to 230 V. The current is alternating.

Mains supply

The electricity supply to a building has three wires:

- Live
- Neutral
- Earth

The **live wire** supplies the current. On this wire, the voltage alternates positive and negative of the neutral wire.

The **neutral wire** returns the current to the power station after it has passed through all the appliances in the building. The voltage on this wire is close to earth voltage, which we take to be 0 V.

The **earth wire** is connected to the earth. Normally, no current flows in this wire. It is there to provide a path to earth when there is a fault. We explain this later.

To help make sure that wiring is correctly installed, the three wires are colour coded:

- **Live** is brown.
- **Neutral** is blue.
- **Earth** is striped green and yellow.

Cables

The conductors in mains cable (as in most electrical cables) are made of copper. This is because copper is one of the best conductors. The copper is in the form of several copper strands twisted together. This makes the cable flexible. It can be bent into the right shapes for running it around the building. Cable used for connecting appliances needs to be even more flexible.

The photo shows how a 3-core mains cable is made up. It has three cores of twisted copper strands for live, neutral, and earth. Each core is surrounded by a layer of plastic to insulate it from the other cores.

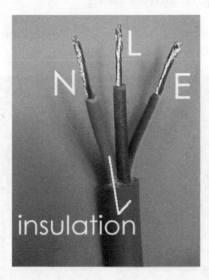

The three insulated cores are surrounded by an outer plastic layer to hold them together and to give more insulation between the cores and anything that the cable is in contact with.

Switches

Most appliances have a switch for turning them on or off. The switch must always be wired into the live wire of the supply to the appliance. In the top diagram (right), the switch is open. No current can flow through the appliance, so it is off. It is still connected to the neutral line of the mains, but this line is close to earth voltage. If there is a fault in the appliance, there is little danger.

In the middle drawing, the switch is closed and the appliance is on. Current flows through the appliance between the live and neutral lines. There is little danger.

In the bottom drawing, the switch is open and the appliance is off. But it is *still connected to the live wire*. Should there be a fault, the appliance is still live and this is dangerous.

The switching drawing shows the earth wire not connected to anything. Connections for the earth wire are described in the next topic.

Self test

1 Why is the core of a cable made from copper?

2 What is the reason for making the core from thin strands twisted together?

3 What are the names of the three wires in an electrical supply? What does each do?

4 Why are the cores of a cable covered with plastic?

5 Why is it important for the plastic to be flexible?

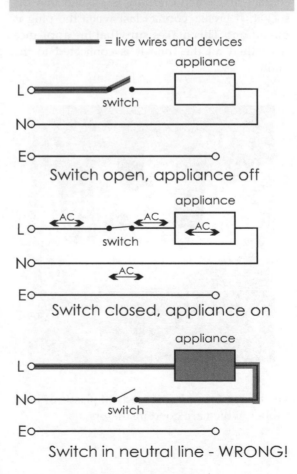

Switch open, appliance off

Switch closed, appliance on

Switch in neutral line - WRONG!

7 Plugs and fuses

Plugs and sockets

Electrical appliances such as electric cookers and immersion heaters need a heavy current. They are usually connected through a switch directly to the mains supply. The supply to ceiling lamps and other fixed appliances is also permanently wired to the mains. A qualified electrician does this.

A moveable appliance is usually plugged in to a socket on the wall. The plug has three pins, live, neutral, and earth. The pins are made of brass. This is a good conductor. It is not quite such a good conductor as copper, but it is stiffer and more hard-wearing. The earth pin is longer than the other two so that it makes contact first when the plug is pushed into the socket. It breaks contact last when the plug is pulled out. This makes sure that the appliance is earthed all the time it is connected to the mains.

The body of the plug is made of plastic or hard rubber, which are good insulators.

Wiring a mains plug

The photograph shows a mains plug with the top removed.

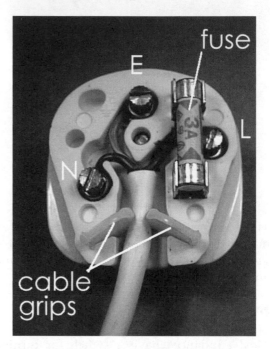

The steps for wiring the plug are:

1   Unscrew the fixing bolt and remove the top from the plug.

2   Look at the instructions supplied with the plug to find out how much of the outer insulation to remove. Use a wire-stripper to remove it.

3   Check whether any of the three wires need to be shortened. If so, cut it shorter.

4   Look at the instructions to find out how much insulation to remove from each wire. Use a wire-stripper to remove it.

5   Twist the bare copper strands with your fingers to make them stay together.

6   Usually it is best to thread the cable through the cable grip at this stage.

7 Loosen the terminal screws on each pin. Wrap the bare cores around each screw and tighten each screw.

8 Make sure that the wires run neatly inside their channels.

9 Check that there no loose strands of copper.

10 Knowing the maximum current that the appliance will require, select a fuse of slightly greater rating. Press it into the fuse clips.

11 **Check again that the three wires are connected to the correct pins.**

12 Replace the top and tighten its retaining screw. If the cord grip has screws, tighten these too.

Fuses

Fuses are used to protect equipment and people against electrical faults. A fuse contains a thin wire made of a special alloy that melts at a fairly low temperature. If the current through the fuse is too high, heat is produced faster than it can be lost from the fuse. The fuse wire gets so hot that it melts and breaks the circuit. This disconnects the supply.

Self test

Fuses are always placed on the live side of an appliance. Why?

(Hint: see drawing on p. 19)

Fuses are rated to blow if the current exceeeds a stated amount. Mains fuses such as the one in the plug (opposite) are made in 3 A, 5 A and 13 A ratings. Always select the rating according to the device being protected. Use 13 A fuses only with high-current appliances such as room heaters and vacuum cleaners.

Danger from electricity

- Touching live metal parts may cause burns or, in serious cases, loss of life. Often, the breathing muscles are paralysed. First aid: turn off the power; then apply artificial respiration.
- A current too heavy for the cable will make the cable hot. This may cause heat damage to the cable and the appliance. It may also lead to fire.
- Sparks, either from static charge (p. 4) or from switches and other devices may ignite inflammable vapours or dusts in the air. This may result in fire or explosion.

Circuit breakers

Some kinds of circuit breaker are used instead of fuses to protect a faulty appliance from drawing too much current. A **thermal circuit breaker** becomes overheated by excess current passing through it. This triggers the breaker switch to open.

An **earth leakage circuit breaker (ELCB)** has a different action. It switches off the mains supply to an appliance when it detects a leakage of current to earth. For example, an exposed metal part of an appliance (such as its case, or a control knob) may become 'live' because of faulty insulation. A person touching that part conducts current to earth and receives a shock. The ELCB measures the current flowing along the live wire and that flowing along the neutral wire. Because some of the current is leaking away (through the person) the live and neutral currents are unequal. The ELCB detects this state and switches off the supply.

A typical ELCB detects a current leak as small as 30 mA, and switches off within 20 ms.

ELCBs do not switch off if the leakage is direct from live to neutral. This is because the currents are *equal*. A person accidentally touching both live and neutral wires or terminals at the same time receives a shock.

8 Electricity in the home

In this Topic, the word 'home' includes the house you live in, the garden, the garage and (if you have one) the workshop and swimming pool. In all of these you may have appliances that run on electricity. Some use mains power, others use batteries. These appliances convert electrical energy into other forms of energy.

Example

An immersion heater converts electrical energy into heat energy.

We use these other forms of energy for doing useful work in the home.

Things to do

Make a list of the electrical appliances in your home. For each one, state into what form of energy it converts electrical energy.

Self test

1 Most appliances produce light, heat or motion. What other form of energy do some appliances produce?

2 Some appliances convert electrical energy into two or more forms. Name three appliances which do this. Which of the forms are useful? Which are wasteful?

Power

The **rate** at which an appliance converts energy is its power. Power is expressed in watts (W). One thousand watts is 1 kilowatt (kW)

Example

A typical electric kettle runs at 2 kW.

Things to do

Go through your list of appliances and find out how many watts each one takes. This information is often marked on a small panel on the back or underside of the appliance. Or you may find it in the manual. Do not forget to include some battery-powered appliances in your list. Examples are an electric torch and a digital clock.

Amounts of energy

The amount of energy converted by an appliance depends on:

• Its power rating.
• The length of time it runs.

The higher the power, the more energy it converts in a given time. The longer the time, the more energy it converts for a given wattage. Putting this as a formula:

$$\text{energy converted} = \text{power} \times \text{time}$$

For mains-powered electrical appliances, it is more practical to rate the power in kilowatts and the time in hours. The unit of electrical energy is the **kilowatt-hour,** symbol **kWh**.

Example

An immersion heater, power 8 kW, takes 3 hours to heat the water in the tank. How much energy does it convert while doing this?

Energy converted = power × time
$$= 8 \times 3$$
$$= 24 \text{ kWh}$$

Things to do

Select three appliances from your list. You already know their power.

Find out the average times they take to perform a task.

Examples

How long does the kettle take to boil enough water for a pot of tea? How long does a vacuum-cleaner take to clean a room? How long does the oven take to cook a joint of meat?

Calculate how much energy they convert while doing these tasks.

Paying for electric power

The amount of electricty used in a home is measured with an electricity meter, connected in to the mains supply line. This reads the amount in kilowatt-hours. The electricity bill charges for this amount of energy at a fixed cost per kilowatt-hour. On the bill, the kilowatt-hour is sometimes referred to as a 'Unit'.

The cost is worked out on the bill using the formula:

cost = number of units × cost per unit

Example

A security floodlight runs at 500 W. It is switched on every night for 10 hours. Electric power is charged at 7 p per kWh. How much does it cost to run the lamp each night?

In kilowatts, the rating of the 500 W lamp is 0.5 kW. Use this value in the formula.

Energy converted = 0.5 × 10 = 5 kWh

Cost = 5 × 7 = 35 p

Things to do

Find out the cost of a kilowatt-hour of electricity from your local supplier.

You have calculated how much energy is used by your three chosen appliances to perform a typical task. Use this information to calculate how much it costs to do this.

Have discussion in class about ways of saving electric power in the home.

Self test

1 A 1000 W microwave oven cooks a bowl of porridge in 4 minutes. If electricity cost 6.93 p per kWh, what is the cost of cooking the porridge?

2 The 2 kW heater of a fan-heater is found to be switched on for 20% of the time. The fan, runs all the time and takes 10 W. At 6.93 p per kWh, what is the daily cost of running the fan-heater for an 8-hour day?

Power supply unit   Extension Box 8

Cells and batteries are ideal for powering hand-held portable equipment such as torches and pocket calculators. Other devices that work on low voltage are powered from the mains, using a low-voltage power supply unit (or **PSU**).

A PSU contains a **transformer** (p. 45) to reduce the voltage from mains voltage to a lower voltage, such as 6 V. It may have a means of switching to a number of standard voltages, such as 3 V, 6 V, 9 V and 12 V. It usually has a **rectifying circuit** (p. 58) to convert the AC output of the transformer into DC. It may also have a **regulator circuit** (p.181), so that it produces a steady voltage even when the load on the PSU varies.

Most PSUs produce only small currents, generally less than 1 A, but this is enough for driving most low-current equipment.

PSUs are made in various forms:

- a plastic body that looks like a large mains plug, with three pins for plugging directly into a mains socket. It has a flexible lead for connecting to the equipment.

- a case with two leads, one to connect to a mains plug and the other low-voltage lead to connect to the equipment.

- a bench PSU is used in laboratories and workshops. It connects to the mains and has terminals that provide the low-voltage DC output. It is usually possible to switch to a range of different voltages and also to vary the output voltage smoothly over a wide range, typically from 0 V to 30 V or more. A limit can be set on the current delivered, to protect circuits powerd by the PSU. Meters display the voltage and current being delivered.

Solar cells   Extension Box 9

A solar cell is a slice of silicon coated on both sides with metallic contacts. The silicon has been processed to give it semiconducting properties.

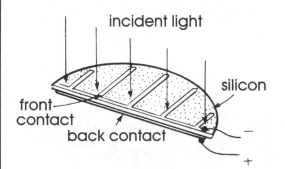

As a result of these properties, a voltage is developed between the front and back contacts when light shines on the front surface of the cell.

A typical solar cell develops 0.45 V in full sunlight, and a current of up to 200 mA. For larger voltages and currents, several solar cells are connected together into a battery, mounted on a panel. A battery of over 60 cells can supply 12 V at 500 mA.

Solar cells are a useful power source for equipment that is remote from the mains supply.

Examples

Microwave transmitters on mountain tops, spacecraft and satellites. In countries such as Western Australia, which has plenty of sunshine, there are solar-powered radio-telephones beside out-of-town highways for emergency calls.

Cells in series Extension Box 10

When we connect two or more cells to make a battery, we normally join the cells **in series**. The cells are connected positive-to-negative, as seen in the drawing below.

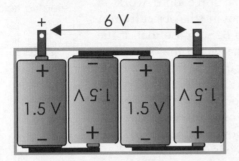

The drawing shows four 1.5 V cells in a battery box. When cells are joined in series like this, the total voltage of the battery is the sum of the voltages:

$$1.5 + 1.5 + 1.5 + 1.5 = 6 \text{ V}$$

Cells in ready-made batteries (p.11) are also joined in series.

Unit of charge Extension Box 11

The unit of electrical charge is the **coulomb**. Its symbol is **C**.

A coulomb is defined as the amount of charge flowing past a point in a circuit when a current of 1 amp flows for 1 second:

$$\text{charge} = \text{current} \times \text{time}$$

Example

A car headlamp takes a current of 450 mA. How much charge passes through the lamp during a 30-minute journey at night?

The current is 0.45 A and the time is 30 min = 1800 s.

$$\text{charge} = 0.45 \times 1800 = 810 \text{ C}$$

Self test

A security siren operates on 300 mA. How much charge passes through the siren when it sounds for 2 minutes?

Root mean square Extension Box 12

One way of stating the size of an alternating current or voltage is to quote its **amplitude** (p. 17). But amplitude tells us the *peak* current or voltage. Sometimes, it is more useful to know the *average* .

Simply taking the average of all values does not work.

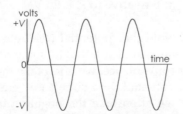

The voltage values of the positive half-cycle are exactly cancelled out by the negative values during the next half-cycle. The average value is *zero*!

To avoid this difficulty, we use the fact that the squares of all numbers are positive.

To find this new kind of average, we use some special maths to square all the thousands of voltage values during a cycle, find the average (or mean) of the squares and finally take its square root. The kind of average we get is called the **root mean square** or **rms**, for short. The maths is too complicated to describe here, but we can quote its result:

The rms value of a sinewave is 0.707 times its amplitude

Putting it the other way round:

The amplitude of a sinewave is 1.4 times its rms value.

Example

When we say that the mains voltage is 230 V, we are quoting its rms voltage. Its amplitude is 1.4 × 230 = 322 V. The voltage peaks to ±322 V and circuit designers must allow for this when choosing components to use.

Joules and watts Box 13

In Physics, we learn about the **joule,** symbol **J.** This is the unit of **work.** Work is done when a force is applied to an object and the object moves in the direction of the force.

Examples

Work is done when the wind blows against a leaf and it moves in the direction of the force.

Work is done when an electric force is applied to an electron and it moves in the direction of the force..

In both these examples we can see that, when work is done, one form of energy is converted into another form. Force is converted into motion. The *amount* of work done, or energy converted, is stated in joules.

The *rate* of doing work (or converting energy) can be stated in joules per second. However, we have a special unit for this, the watt.

1 watt = 1 joule per second.

Example

A man does 2800 J of work to climb up a ladder. He takes 5 seconds to climb. What is his rate of conversion of energy.

He applies 2800 J in 5 s.
His rate of working is 2800/5 = 560 J/s
In watts, this is 560 W.

Self test

1 An electric motor does 5 J of work while moving a toy car 1 metre in 2 s. What is its rate of conversion of energy?

2 How much work is done by a 25 W lamp, switched on for 1 minute?

Volts Extension Box 14

Imagine a positively charged object in an electric field. A force *F* is applied to the object to move it from point A to point B, against the force of the electric field.

The force does work in moving the object from A to B. Suppose that, if the charge on the object is 1 coulomb, the work done is 20 J. If the charge is doubled to 2 C, 40 J of work are required. To move a charge of 3 C needs 60 J.

When the charge is at B, it has **potential energy,** stored in it by moving it from A to B.

If the force is removed, the object is repelled by the field and moves back to A. As it does so, it loses potential energy and gains energy of motion. Its potential energy at point B is equal to the work that must be done to move it to B. This we have seen is 20 joules per coulomb. Whatever charge we move from A to B takes 20 J/C. This is the **electrical potential** at B relative to A.

We have a special name for the unit of potential, instead of J/C. We call it a **volt,** symbol **V.** So, in this example, the potential at B relative to A is 20 V.

Note that potential (or 'voltage') is always measured relative to some other point. For convenience, we often take the Earth as zero potential. In a circuit powered by a battery we often take the negative terminal of the battery as 0 V.

Self test

If it takes 56 J to move an object charged with 4 C from A to B, what is the potential of A relative to B? Of B relative to A?

Questions about electricity

(Answers to questions with short answers are given at the back of the book)

1 Describe how you would show that like charges repel each other.

2 Name three substances that can be charged by rubbing them with a woollen cloth.

3 What are the names of the two kinds of electric charge. Which kind of charge is carried on an electron?

4 Why can we not charge copper by rubbing it with a cloth?

5 Why is copper such a good conductor of electricity?

6 What causes lightning?

7 List four   good conductors and four insulators.

8 What is an electric current?

9 What is the difference between a cell and a battery?

10 List two types of cell, describe their features and give examples of their uses.

11 Name a type of rechargeable cell, describe its features and give examples of its uses.

12 What types of battery could you use for backing up the power supply to a security system? If the system is to run on 24 V, how many cells of each type would you require?

13 In what ways may a 12 V car battery be dangerous?

14 Name the units of: (a) current, (b) voltage, and (c) power. What are their symbols?

15 Write an equation that shows how the three units of Question 14 are related to each other.

16 What are the units and symbols for (a) a millionth of a volt, and (b) a million watts?

17 A halogen lamp runs on 12 V with a power of 20 W. How much current does it take?

18 A small radio receiver uses two alkaline cells and takes a current of 12 mA. What is its power?

19 Draw a diagram to illustrate 4 cycles of AC. Label the amplitude and the period. If the period is 20 ms what is the frequency?

20 Name the three lines of the mains supply and explain what they do.

21 Explain why a power switch must be placed on the live side of an appliance.

22 Why should an ELCB be used with an appliance such as an electric lawn-mower?

23 An electric kettle rated at 2.2 kW takes 5 min to bring water to the boil. How much does this cost if electricity is charged at 7.2 p per unit?

Extension questions

24 Describe what happens when sodium chloride ionises in solution.

25 What conductor is often used in power lines?

26 What is conventional current?

27 How does a neon lamp work?

28 Describe some of the dangers of static electricity.

29 Explain the meanings of these terms: electrolyte, cathode, anode.

30 In a photocopier, why does the black powder form a positive image on the drum?

31 In a photocopier, what are the corona wires and what do they do?

32 Describe a solar cell and list some ways in which solar cells are used.

33 A sinewave produced by an audio generator has an amplitude of 6 V. What is its rms value?

34 Draw a diagram showing three 1.5 V cells connected in series. What is the total voltage?

35 An electric torch has four 1.5 cells in series. It has a 0.96 W lamp. How much charge passes through the lamp while it is switched on for 10 min?

36 A krypton lamp takes 0.7 A when running on a 2.4 V supply. How much work is done while the lamp is switched on for 20 min?

37 How much work is done in moving a charge of 25 C from one terminal of a 6 V battery to the other?

9 Resistance

We begin by investigating what happens when we use different voltages to make a current flow through a conductor. We use a poor conductor, such as carbon, so that the current will not be too large to measure with an ordinary multimeter.

Things to do

A suitable piece of carbon for this investigation is a short length (about 3 cm) of Artists' charcoal .

1 Connect a 6 V battery, a multimeter and the piece of carbon in a circuit, as shown in the photograph. Use short leads with crocodile clips at each end.

2 Leave the positive (red) probe of the meter free. By touching this against different metal contacts in the battery box, you can obtain voltages of 1.5 V, 3 V, 4.5 V and 6 V. Alternatively, use a PSU with switchable output instead of the battery.

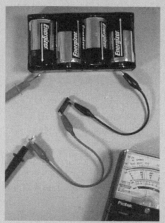

3 Set   the multimeter to its largest current range (500 mA or 1 A). If the readings are too small to read accurately, you can switch to a lower current range.

4 Take the first reading with the red probe at the 6 V point. Measure the current, using the meter. Record these results in a table:

Voltage (volts)	Current (amps)	voltage/ current
6		
4.5		
3		
1.5		

5 Repeat with the voltage set to 4.5 V, 3 V and 1.5 V.

What do you notice about the current as we reduce the voltage? To investigate this further, divide each voltage by the corresponding current and write the results in the third column. What do you notice about the values in the third column?

Ohm's Law

The results of the investigation above show that:
> **The current passing through the carbon is proportional to the voltage difference between its ends.**

This was first discovered, using lengths of wire, by Gregor Ohm, so it is called **Ohm's Law**. It applies to all conductors. We can state it as an equation:

$$\frac{voltage}{current} = \text{constant}$$

You may have noticed in your class that different people, working with different pieces of carbon have found different constants. The value of the constant is a property of a particular piece of carbon. We call it the **resistance** of the piece of carbon. Now we can write the equation like this:

$$\frac{voltage}{current} = \text{resistance}$$

Units of resistance

If the voltage is expressed in volts and the current in amps, the unit of resistance is the **Ohm**, symbol Ω. This symbol is the Greek capital letter *omega*.

Self Test

1 When a voltage of 6 V is applied to a piece of carbon, the current is 120 mA. What is the resistance of the carbon?

2 When a voltage of 230 V is applied to a lamp, the current is 260 mA. What is the resistance of the lamp?

Larger units of resistance are the kilohm (k) and megohm (M).

$$1 \text{ k}\Omega = 1000 \text{ }\Omega$$
$$1 \text{ M}\Omega = 1000 \text{ k}\Omega$$

Ohm's Law equations

The equation given opposite can be written in three forms:

$$\text{resistance} = \frac{\text{voltage}}{\text{current}}$$

$$current = \frac{voltage}{\text{resistance}}$$

$$voltage = current \times \text{resistance}$$

If we are given any two of the quantities, we can calculate the third. These three equations are used more than any in electronics, so you will need to remember them. An easy way of remembering is to memorize this diagram:

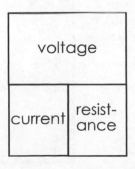

To use the diagram, cover the quantity that you want to calculate. The diagram shows the remaining quantities as they appear in the equation.

Example

To calculate current:

Current equals 'voltage over resistance'.

Try it for the other two quantities.

Self test

1 When 25 V is applied to a length of cable, the current through it is 10 A. What is the resistance of the cable?
2 A coil of wire has a resistance of 600 Ω and the voltage across it is 4.5 V. How much current flows through the cable?
3 If a wire has a resistance of 2.4 Ω, and a current of 3.5 A is flowing through it, what is the voltage difference between its two ends?
4 A 2.3 kW heater runs on the 230 V mains. What is its resistance?

Symbols for quantities

You have already used symbols for *units*, such as A, V, and W. It is useful to have symbols for *quantities*, too. This makes it quicker to write out equations. The symbols for quanties are:

$$I \text{ for numbers of amps}$$
$$V \text{ for numbers of volts}$$
$$R \text{ for numbers of ohms}$$

Symbols for quantities are in slanting letters (italics). Using these symbols instead of words, the Ohm's Law equations become:

$$R = V/I$$
$$I = V/R$$
$$V = IR$$

Try not to confuse V, which means 'volts' with *V*, which means 'numbers of volts'

10 Resistors

In most circuits, we join the components together with copper wires. This is because copper is a good conductor of electricity. It has very low resistance.

Some connections may need greater resistance than that of a copper wire. This is when we use resistors.

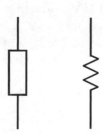

The photo shows a typical **fixed resistor**. Resistors of this kind are sold in a range of different resistances, from less than 1 Ω and up to 10 MΩ.

These are two different symbols used for resistors in circuit diagrams. We use the rectangle symbol in this book. The zigzag symbol is less often used nowadays.

Preferred values

Resistors are made in a range of values (in ohms):
1.0 1.1 1.2 1.3 1.5 1.6 1.8 2.0 2.2 2.4 2.7 3.0
3.3 3.6 3.9 4.3 4.7 5.1 5.6 6.2 6.8 7.5 8.2 9.1

After these 24 values, the sequence repeats in multiples of ten:

10 11 12 13 ... up to ... 82 91,
then 100 110 120 ... up to 820 910,
then 1 k 1.1 k 1.2 k ... up to 8.2 k 9.1 k,
('k' means kilohms)
then 10 k 11 k 12 k ... up to 82 k 91 k,
then 100 k 110 k 120 k ... up to 910 k 1M.

This is the **E24 series**.

Resistor colour code

Three coloured bands are used to tell us the resistance of a fixed resistor. The bands are close together at one end of the resistor. The colour of each band represents a number.

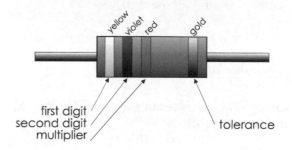

first digit
second digit
multiplier

tolerance

Reading from the end, the meaning of the bands are:

First band — **First** digit of resistance
Second band — **Second** digit of resistance
Third band — **Multiplier** — a power of 10, or the number of zeros to follow the two digits.

The table shows the meanings of the colours.

Example 1

The bands are yellow, violet, red.

Yellow means '4'
Violet means '7'
Red means '2'

Write '4', then '7', then follow with two zeros. This gives:
4700 Ω

We normally write this as:
4.7 kΩ

Colour	Number
Black	0
Brown	1
Red	2
Orange	3
Yellow	4
Green	5
Blue	6
Violet	7
Grey	8
White	9

Example 2

The bands are white, brown, yellow.

White means '9'
Brown means '1'
Yellow means '4'

Write '9' then '1', then follow with four zeros.
This gives:

$$910\,000\ \Omega\quad\text{or}\quad 910\ \text{k}\Omega$$

There is a similar system with four bands for marking high-precision resistors.

Self test

1 State the resistances of resistors that have bands of these colours:
(a) orange, white and brown.
(b) green, blue and yellow.
(c) brown, black and green.
(d) brown, black and black.
(e) red, red, and red.

2 What colours are the bands on resistors of these values?
(a) 33 Ω, (b) 200 kΩ, (c) 750 Ω, (d) 43 kΩ, (e) 820 Ω.

Breadboards

A breadboard makes it easy and quick to build circuits. It is a plastic block with rows of sockets. The sockets in each row are connected electrically, as shown in the photo.

If you plug two or more component wires into the same row, current can flow from one to the others.

Things to do

You need:
- 10 fixed resistors of different values.
- A multimeter.
- A breadboard.

1 Find out the value of one of the resistors by reading its colour code.
2 Use the multimeter to check the resistance. First select a suitable range on the meter.
3 Plug the wire leads of the resistor into *different rows* on the breadboard. Touch one probe to each of the resistor leads.

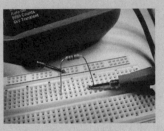

4 Read the resistance on the meter. Does it agree with the colour code?
5 Repeat with the other resistors.

Resistance on an analogue meter

The resistance scale runs from right to left and is not linear. This means that the markings are more crowded toward the higher (left) end of the scale.

The reading in the photo (on the top scale) is 72 Ω.

When using an analogue meter, adjust the zero setting occasionally. To do this, touch the two probes together. Then set the 'Ohms adjust' knob or wheel until the needle lines up with zero (on the *right* of the scale).

11 More about resistors

If we want to be able to vary the resistance of part of a circuit, we use a **variable resistor**. One type of variable resistor is called a **potentiometer**. This type is a often used for volume controls in audio equipment. It is more often called a 'pot', for short.

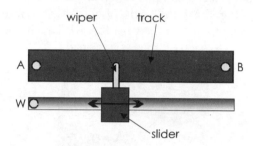

Above is a slider pot, which has a track made from a film of carbon. In the more expensive, hard-wearing and accurate types, the track is a layer of conductive ceramic. The ends of the track are connected to two terminals.

The third terminal is connected to the **wiper.** This is a springy metal strip that presses firmly against the track and makes electrical contact with it. It is attached to a sliding knob, used to move the wiper along the track. As it moves, the distance between one end of the track (say, A) and the wiper (W) is changed. This changes the electrical resistance between A and W. The resistance can have any value between zero and the resistance of the whole track. Slider pots are often used on audio equipment (p. 165) for setting the frequency response.

A common form of variable pot is the **rotary pot**. This has a curved track, along which the wiper moves as the shaft is turned.

Often the track covers an angle of about 270°.

Rotary pots are often used as volume controls in audio equipment and also for controlling the brightness of lamps, the speed of motors and many other purposes. The shaft can be fitted with a knob, if preferred.

Sometimes we want to adjust a resistance when we are setting up a circuit. After 'pre-setting' or 'trimming' the circuit, we may not need to change the resistance again.

We use a smaller rotary pot, known as a **preset pot**, or as a **trimmer**, or **trimpot**, which is adjusted with a screwdriver.

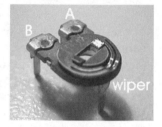

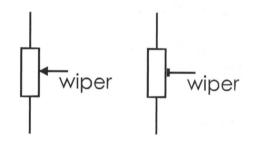

Special symbols used for pots (left) and presets (right). Zigzag versions are also used.

Design tip

Pots need careful use. If the wiper is near one end of the track, the resistance between the wiper and the near end is small. As a result, the current through that part of the track is too high and burns out the track. If possible, wire a small fixed resistor in series with the wiper or track to limit the current to a safe amount.

Power ratings

Many fixed resistors are intended to conduct electricity at a maximum power of a quarter of a watt (250 mW). This power must not be exceeded.

Example 1

The current through a 250 mW resistor is 10 mA and the voltage across it is 5 V. The power is 0.01 × 5 = 0.05 W = 50 mW. Because the resistor is rated at 250 mW, it can safely run at 50 mW. It becomes only slightly warm as a result of the current passing through it..

Example 2

If the current through a 250 mW resistor is 50 mA and voltage is 5 V, the power is 250 mW. This is as much as the resistor can safely stand. It will get hot but it is not damaged. If either the current or voltage is increased above these amounts, the resistor gets too hot. It may scorch or burn and possibly crack into pieces. Even if it is not totally destroyed, its resistance changes permanently as a result of overheating.

Resistors are made with higher power ratings, such as 0.5 A, 1 A, 5 A. Some are able to run at several hundred watts. These are larger than the typical low-power resistors. Those of the highest ratings usually consist of a coil of thin wire wound on a ceramic core.

Typical pots are rated to run at a maximum of 0.2 W to 0.5 W. Pots of higher rating, up to 3 W, usually consist of a coil of wire wrapped on a ceramic core.

Tolerance

There is usually a fourth coloured band on a resistor, at the end opposite to the three bands. This band indicates the tolerance or precision of the resistor. This tells us how far the actual resistance may differ from the **nominal resistance**, as shown by the colour code

Example 1

A 470 Ω resistor has a gold tolerance band. This means a tolerance of ±5%. Calculate 5% of 470Ω, which is 470 × 5/100 = 23.5Ω.

Colour	Tolerance
Red	±2%
Gold	±5%
Silver	±10%
No band	±20%

The actual resistance of the resistor may be somewhere between:

$$470 - 23.5 = 446.5 \ \Omega$$
and $\qquad 470 + 23.5 = 493.5 \ \Omega$

Example 2

A 220 kΩ resistor has no tolerance band. Its tolerance is ±20%. Calculate 20% of 220 kΩ, which is 220 × 20/100 = 44 kΩ.

The actual value of the resistor may be somewhere between:

$$220 - 44 = 176 \ k\Omega$$
and $\qquad 220 + 44 = 264 \ k\Omega$

The reason for E24

It would be very expensive to make and stock all possible values of resistor from 1 Ω up to 1 MΩ. Also, because 5% tolerance is good enough in most circuits, the limited number of resistances of the E24 series (p. 30) covers the whole range of values. Take these four nominal values as examples:

Nominal	Lowest (–5%)	Highest (+5%)
390	370.5	409.5
430	408.5	451.5
470	446.5	493.5
510	484.5	535.5

The range of each value *slightly* overlaps each of its neighbours. At 5% tolerance, there is no point in making resistors with values in between these E24 values.

12 Resistor networks

When two or more resistors are joined together we create a **resistor network**. This Topic describes several types of network and their properties.

Resistors in series

If two or more resistors are joined end-to-end so that current flows through each one in turn, we say they are joined **in series**. Compare this with cells joined in series, p. 25. We find the **effective resistance** of the series by **adding** together their resistances.

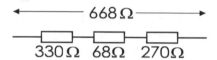

Example

In the series of three resistors above, the effective resistance is 330 + 68 + 270 = 668 Ω. As a check on the calculation, note that the effective resistance is always bigger than the biggest of the individual resistances.

Self test

1 What are the effective resistances of each of the following groups of resistors when wired in series?
(a) 220 Ω and 3.3 kΩ.
(b) 5.6 Ω, 39 Ω, and 16 Ω.
(c) 13 kΩ , 1 MΩ and 390 kΩ.

2 A student needs a 470 Ω resistor, but the lab is out of stock. How could this be made up from resistors of other E24 values?

3 Make up a 130 kΩ resistor from other E24 values.

Things to do

You need:
- 10 fixed resistors of different values and various tolerances.
- A multimeter.
- A breadboard

1 Read the colour codes to find the resistance of each resistor. Then check your reading by measuring the resistances with a multimeter (p. 31).

2 Take any two or three resistors and plug them into the breadboard so they are in series. Calculate the effective resistance of the series. Then check your calculations by using the multimeter to measure the resistance of the series. Repeat for four more groups of resistors.

3 From its colour bands, find the resistance and tolerance of a resistor. Work out what its lowest and highest resistances could be. Check with a multimeter that the actual resistance is within these limits.

Current rules

1 Three or more wires in a network meet at one point. It is not possible for charge to build up at the junction. It is not possible for charge to decrease at the junction. Therefore:
> **The total current arriving at a junction equals the total current leaving it.**

Example

The total current arriving is 2.1 + 2.4 = 4.5 A
The total current leaving is 1.5 + 3.0 = 4.5 A

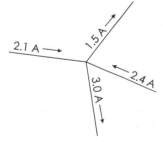

2 In a series circuit (one in which all the components are in series), there is no point at which charge can enter or leave the circuit. Therefore:

The current is the same everywhere in a series circuit.

Example

The cell and three resistors are in series. The same current (= i amps) flows in all parts of the circuit.

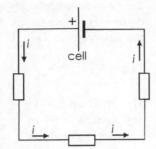

Voltage rules

1 Going round a series circuit in the direction of the current, there is a voltage drop across each of the resistors. There is a voltage rise across the cell. The size of each voltage drop depends on Ohm's Law. The voltage rule is that:

The sum of the voltage drops in a series circuit equals the voltage across the circuit.

Example

The voltage drops are v_1, v_2, and v_3. Their total is equal to the voltage *rise* v_T across the cell.

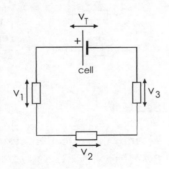

2 The drawing shows a circuit in which the resistors are joined **in parallel.**

One end of each resistor is connected to the positive terminal of the cell. The other end of each resistor is connected to the negative terminal. This means that:

In a parallel circuit, there is the same voltage across each component.

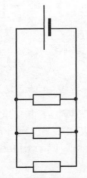

Voltage divider

This kind of network is also called a **potential divider**. The input to a voltage divider is a voltage v_{IN}. This drives a current i through the two resistors. Because the two resistors are in series, the same current flows through both (Current Rule 2).

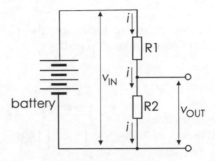

The effective resistance of the two resistors in series is $R_1 + R_2$. The voltage across them is v_{IN}. By Ohm's Law, the current is:

$$i = \frac{v_{IN}}{R_1 + R_2}$$

Using Ohm's Law again, the voltage across R2 is:

$$v_{OUT} = i \times R_2$$

Substituting the value of i from the previous equation:

$$v_{OUT} = v_{IN} \times \frac{R_2}{R_1 + R_2}$$

This is the equation for calculating the output voltage of a voltage divider. By choosing two suitable resistors, we can obtain any output in the range from 0 V to v_{IN}.

Example

In a voltage divider, $v_{IN} = 6$ V, $R_1 = 220 \ \Omega$ and $R_2 = 390 \ \Omega$. Calculate v_{OUT}.

$$v_{OUT} = 6 \times \frac{390}{390 + 220} = 3.84 \text{ V}$$

There are questions about the current rules, the voltage rules and about voltage dividers on page 39.

Resistances in parallel Box 15

The effective resistance of two or more resistances in parallel (p. 35) is calculated from:

$$\frac{1}{R} = \frac{1}{R_1} + \frac{1}{R_2} + \frac{1}{R_3} + \dots$$

There are as many terms on the right as there are resistances in parallel.

Example

The effective resistance of this network is found by calculating:

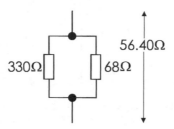

$$\frac{1}{R} = \frac{1}{330} + \frac{1}{68} = 0.00303 + 0.01471 = 0.01774$$

$$\Rightarrow R = \frac{1}{0.01774} = 56.40 \ \Omega$$

Check: the effective resistance is always *less than* the smallest of the resistances in parallel.

Self test

Find the effective resistance of each of the following sets of resistors, when wired in parallel:

(a) 22 Ω and 270 Ω, **(b)** 120 Ω and 10 kΩ, **(c)** 27 Ω, 47 Ω, and 15 Ω, **(d)** 390 kΩ, 18 kΩ, and 91 kΩ, **(e)** 100 Ω, and 100 Ω.

Things to do

Repeat the tests you made on p. 34, but with the resistors in parallel.

Mixed networks Box 16

A mixed network has some resistances in series and others in parallel. Look for groups of resistances in the network, which are all in series or all in parallel. Re-draw the network, replacing each group with its equivalent resistance. Gradually simplify the network to a single resistance.

Example 1

The 56 Ω and 33 Ω resistances in parallel can be replaced with 20.8 Ω. This gives two resistances in series. Replace these with 67.8 Ω.

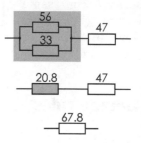

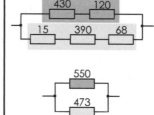

Example 2

There are two sets of resistances in series. Simplify them to 550 Ω and 473 Ω. These are in parallel, so may be replaced with 254 Ω.

Self test

1 Find the effective resistances of these networks:

(a)

(b)

2 What is the effective resistance if the 1200 Ω of the network above is **(a)** removed, and **(b)** replaced with a copper wire?

Resistor printed code Box 17

This is sometimes used instead of the resistor colour code. It is also known as the **BS1852 code.** The code uses one of three letters to show the unit and the position of the decimal point. The letters are:

R = ohms,   **K** = kilohms,   **M** = megohms.

Examples

Code	Value	Code	Value
8R2	8.2 Ω	82K	82 kΩ
82R	82 Ω	820K	820 kΩ
820R	820 Ω	8M2	8.2 MΩ
8K2	8.2 kΩ	-	-

A second letter at the end of the code indicates tolerance:

G = ±2%   **K** = ±10%
J = ±5%   **L** = ±20%

Examples

9K1J means a 9.1 kΩ resistor with ±5% tolerance.
12KK means a 12 kΩ resistor with ±10% tolerance.

Self test

1 What are the values of resistors marked **(a)** 33K, **(b)** 4M7, **(c)** 2R2, **(d)** 1R8J, and **(e)** 27KK?

2 Write the codes for resistors with the following ratings: **(a)** 47 Ω, **(b)** 100 kΩ, **(c)** 9.1 kΩ, **(d)** 3.9 kΩ ±5%, and **(e)** 750 kΩ ±10%.

Multiturn trimmers   Box 18

These are used when we need to be able to adjust a variable resistor with high precision. The wiper is moved by a screw mechanism. It takes several turns of the screwdriver to move the wiper the full length of the track.

The screw head can be seen on the left. It requires 10 turns of the screwdriver to move the wiper from one end of the track to the other. Even more precise trimmers need 18 or 25 turns.

Self test

Can you read the value and tolerance of the trimmer illustrated above?

Energy transfer   Box 19

A charge carrier (such as an electron) loses energy as it moves through a conductor. The energy is transferred to the atoms of the conductor. This makes them vibrate — the conductor becomes hot. Electrical energy is converted to heat. We use this action to heat electric kettles and room heaters. In electronic circuits the resistors, and other components become warm or hot. This is usually a disadvantage because energy is being wasted. In an electric lamp, the conductor is a thin wire that becomes white hot and emits visible light.

The rate of energy conversion is measured in watts (p. 13).

More about voltage dividers

Things to do

You need:
- A breadboard.
- A power supply (6 V battery or PSU).
- A 22 kΩ resistor, a 10 kΩ resistor, and one other.
- A multimeter.

1 With the 22 k resistor and one other, build a voltage divider to give an output of 3.6 V, as shown in the drawing. You have to work out what the value of the other resistor (R2) should be. Do not put the 10 kΩ resistor in the circuit yet.

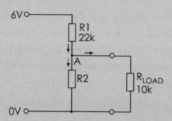

2 Use the multimeter to check that the divider is working. Its input voltage should be 6 V and its output should be 3.6 V.

3 Connect the load resistor (10 kΩ) to the divider. Measure the output voltage again. What has happened to it?

The results of the investigation above are explained by thinking of the currents at point A. According to current rule 1 (p. 34) the currents leaving point A to flow through R2 and the load, must be equal to the current arriving through R1. The current splits and more goes through the load than though R2. This is because the load has a lower resistance than R2.

Because the current through R2 is reduced when the load is connected, the voltage across R2 is decreased (Ohm's Law). The output of the divider is less than we first calculated. Try working out the voltages and current in the network to check your results.

Circuit diagrams

There are a two things about the circuit diagrams that you should notice:
- **Cell or battery symbol:** These are not often used. Instead, the diagram shows a pair of terminals. One, labelled 0V, is the negative terminal of the power supply. The other is labelled with the voltage of the positive terminal of the supply. The supply *may* be a cell or battery, but it is more likely that you will obtain the supply from a bench PSU.
- **Ohms symbol:** This is omitted. Instead we use a shorter form, using the resistor printed code (p. 37).

Design point

We need to avoid a serious voltage drop caused by the load taking too much current from the divider. A simple rule is that the current flowing through the divider from positive supply to the 0 V line must be at least **10 times** the current being taken by the load. There will still be a drop, but it will not be unduly large.

Variable voltage divider

If a variable output voltage is needed, use a divider based on a variable resistor.

The values of R1 and R2 decide the upper and lower limits of v_{OUT}. Without R1 and R2, the divider output ranges from 0 V to v_{IN}.

Measuring resistance   Box 21

This is another way of measuring resistance. It uses two separate meters, an ammeter for current (I) and a voltmeter for voltage (V).

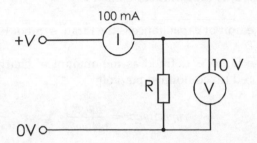

R is the resistance being measured. The ammeter measures the current *through* R. The voltmeter measures the voltage *across* R. The values beside each meter in the diagram are their **full scale deflection**. This is the largest current or voltage that the meter can measure.

The supply $V+$ is variable up to 10 V. Use a PSU or connect different numbers of cells. Try a 4 or 5 different voltages. Measure I and V at each and calculate $R = V/I$ at each.

There is an error in this technique. Some of the current that goes through the ammeter goes *through the voltmeter*, not through R. So the ammeter reading is too big. But a voltmeter takes very little current compared with that through R, so the error is small.

Questions on the current and voltage rules

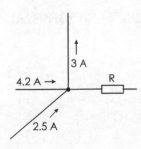

1 How much current flows through R? What is its direction?

2 If the 4.2 A current is reduced to 0.7 A, what is the current through R? What is its direction?

3 In the network on the right, which resistor has the biggest voltage across it?

4 What is the current through the network on the right? What is the voltage drop across each resistor?

5 In the network on the right, what supply voltage produces a current of 30 mA?

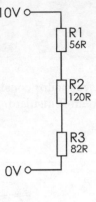

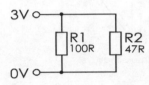

6 On the left, what is the voltage across and the current through each resistor?

Questions on voltage dividers

1 What is the output voltage of the divider on the right?

2 What is its output voltage if the input voltage is raised to 15 V?

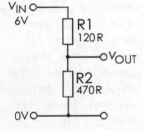

3 What is the output in the diagram above if R1 is increased to 680 Ω?

4 What input is required in the diagram above to give an output of 5 V?

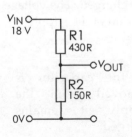

5 What is the output of the divider on the left?

6 What is the output if the input is reduced to 3 V?

7 What is the output if the resistors are exchanged?

8 Design a divider to produce an output of 4.5 V, given an input of 9 V.

9 Design a divider to produce an output of 4.8 V, given an input of 12 V.

13 Capacitors

A capacitor consists of two metal plates with a layer of insulator between them.

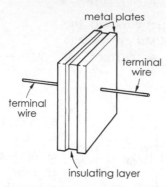

metal plates

terminal wire

terminal wire

insulating layer

The insulating layer may be a thin sheet of plastic, but some types of capacitor have a layer of air instead.

If a capacitor is connected to a source of DC electric power, electrons accumulate on the plate that is connected to the negative supply terminal. These repel electrons from the opposite plate. The repelled electrons flow toward the positive terminal.

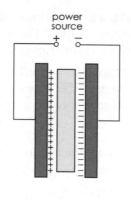

power source
+   −

A capacitor connected like this to a power supply instantly becomes charged. The voltage between its plates equals that of the supply.

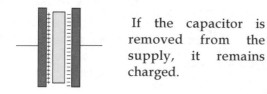

If the capacitor is removed from the supply, it remains charged.

Because of the insulating layer, current can not flow through the capacitor. The capacitor remains charged indefinitely. For this reason, capacitors are useful for **storing charge**.

Capacitance

The ability of a capacitor to store charge is called its **capacitance**, symbol C.

The unit of capacitance is the **farad**, symbol **F**.

The farad is defined as the amount of charge stored (in coulombs) per volt:

$$capacitance = \frac{charge}{voltage}$$

Example

The amount of charge stored on a capacitor is 6 coulombs. The voltage between its plates is 2 V. What is its capacitance?

capacitance = 6/2 = 3 F

Self test

1 A capacitor receives a charge of 2.5 C and the voltage across it is 10 V. What is its capacitance?
2 A 2 F capacitor has 5 V across it. How much charge is it storing?

Capacitors rated in farads are used for backing up the power supply to computer memory. However, most electronic circuits need much smaller capacitances. The units in which most capacitors are rated are:
- **microfarad**, millionths of a farad, symbol **μF**.
- **nanofarad**, thousandth of a microfarad, symbol **nF**.
- **picofarad**, thousandth of a nanofarad, symbol **pF**.

Self test

Express in nanofarads: **(a)** 1000 pF, **(b)** 2.2 μF, **(c)** 1 F, **(d)** 47 pF, **(e)** 56 μF.

Types of capacitor

There are many types of capacitor, of which only the most often used are described here:

Polyester: The insulating material is polyester which gives a relatively high capacitance. The plates are made of metal foil, or a metallized film is deposited on the insulator. The 'sandwich' of plates and insulator is rolled to make it more compact and the roll is coated with insulating plastic. Polyester capacitors (the two on the left below) are general-purpose capacitors and widely used.

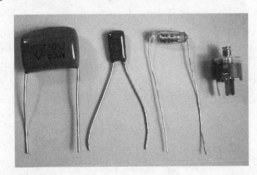

Polystyrene: (third from left above) These are constructed in a way similar to polyester capacitors. Using polystyrene as an insulator results in lower capacitances than with polyester. However, they can be made with much closer tolerance, so are suitable for tuning circuits and filters.

Variable: These have two sets of plates, the alternate plates being electrically connected. One set is fixed. The other set can be turned so that we can vary the amount by which the sets overlap (below). This varies the capacitance.

The larger capacitors used for tuning radio receivers have air as the insulator. Plastic film is used in small trimmer capacitors (right above). Some trimmers work by having a screw adjustment that compresses or loosens the plates and film to vary the capacitance.

Electrolytic capacitors

These are used to store large amounts of charge. Their capacity is usually 1µF or more and may be as much as 10 000 µF.

The two commonest types of electrolytics are **aluminium electrolytics** (left and centre below) and **tantalum bead capacitors** (right).

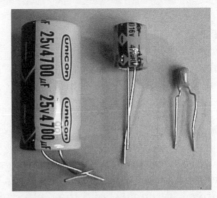

Electrolytics can store large amounts of charge for hours. When building or testing circuits, there is danger from electric shock if you touch the terminal wires without giving the capacitor time to discharge. When storing large electrolytics, twist their terminal wires together (left above) so that they can not retain a charge.

Electrolytics are **polarised**, which means that they have positive and negative terminals. They *must* be connected the right way round. In the photo, there are markings on the case to indicate the negative terminal. If an aluminium electrolytic capacitor is connected the wrong way round, gas is formed inside and it will explode. Tantalum bead capacitors, can be destroyed in a few seconds by wrong connection.

The insulation between the plates in electrolytics (especially the aluminium type) is not as high as that in other types of capacitor., A few microamps leaks across between the plates.

Electrolytics have wide tolerance normally ±20% or wider. They can not be used in precision filters or timing circuits.

Tantalum bead capacitors are made with smaller capacitances than aluminium electrolytics. However, they are generally smaller in size so are useful where space is limited.

14 Charging capacitors

Storing charge

The ability of a capacitor to store charge is one of its most important properties.

Things to do

Set up the circuit shown in the diagram, using a breadboard. The wire from the positive terminal of the electrolytic capacitor is a **flying lead**. It is a piece of wire about 10 cm long. One end is plugged into the same row of contacts as the positive wire of the capacitor. The other can be plugged into the board to connect either to (A) the positive supply or (B) the lamp.

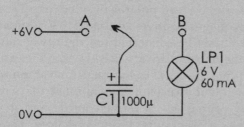

1
Connect the flying lead C to the positive power line (A) to charge the capacitor.

2 Quickly connect C to the lamp at B, to discharge the capacitor. Did you see the lamp flash? If not, try again.

3 Connect the lead to A again. Remove it from the socket, but wait 10 seconds before connecting it to the lamp. Does the lamp flash?

4 Repeat **3**, but wait for longer times before connecting to the lamp. How long does the capacitor hold enough charge to flash the lamp?

This investigation demonstrates that capacitors can hold charge for a long time.

Charging and discharging

The rate of flow of charge depends on the voltage across the capacitor.

Things to do

This circuit has a switch for charging and discharging. It has a resistor to make the current smaller. Because of this, charging and discharging takes longer. It gives you time to see what happens. The meters can be a separate voltmeter and ammeter, or a pair of multimeters.

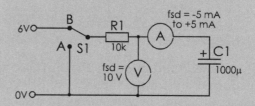

1 Set the switch to A to discharge the capacitor if it is already charged.
2 Set the switch to B and watch the meters as the capacitor is charged. You and a partner could watch one meter each.
3 Set the switch to A and watch the meters as the capacitor is discharged.
4 Repeat **2** and **3** until you can answer these questions:

- When does the biggest current flow into the capacitor?
- When does the biggest current flow out of the capacitor?
- When does no current flow into or out of the capacitor?
- When does the voltage across the capacitor change most quickly?
- When does the current into or out of the capacitor change most rapidly?

In the investigation, the flow is too fast to see exactly what happens. The voltage change can be more clearly seen by connecting an oscilloscope in place of the voltmeter. The display looks something like this:

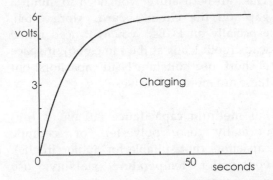

When the switch is turned to A, the voltage on the supply side of the resistor is 6 V and that on the capacitor side is 0 V. By Ohm's Law, the current through the resistor is 6/10 000 = 600 μA. Charging begins and the voltage across the capacitor (see graph) rises steeply. The voltage on the supply end of R1 stays at 6 V, but the voltage at its other end in increasing. The voltage difference across R1 is *decreasing*. Ohm's Law still applies, so the current through R1 is decreasing. This means that the rate of charge of C1 is decreasing and the voltage across it rises more slowly.

The voltage rises more and more slowly until C1 is charged to 6 V. Then there is NO voltage difference across R1 and therefore NO current flows through it. The graph levels out. The capacitor is fully charged.

A curve shaped like the graph above is called an **exponential curve.**

The reverse happens when the capacitor is discharged. At first, there is a voltage difference of 6 V across R1, so 600 μA flows out of the capacitor, through R1 to the 0 V line. The voltage becomes less as the capacitor discharges. The voltage drops more and more slowly. When it reaches zero, the capacitor is uncharged.

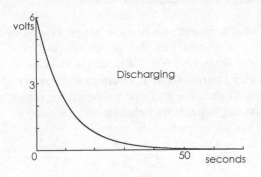

Capacitors in parallel

Wiring two or more capacitors in parallel is the equivalent of adding together the areas of their plates. For this reason, the effective capacitance is the sum of the individual capacitances.

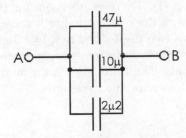

Example

In the diagram above, the effective capacitance is:

$$C = 47 + 10 + 2.2 = 59.2\ \mu F$$

The effective series capacitance is always greater than the greatest of the individual capacitances.

Self test

1 If the supply voltage in the circuit in column 2, p. 42 was increased to 10 V and R1 was increased to 18 kΩ, what would be the current into C1 when the switch was first set to B? What would be the final current into C1?

2 If C1 was decreased to 470 μF, would the time to charge the capacitor be longer? or the same? or shorter?

43

Design tip — Capacitor values

Capacitors are made in a range of values that is similar to the range of preferred resistor values (p. 30). Capacitors have wider tolerances than resistors so there is no point in having 24 basic values in the range. Instead, capacitors values are:

1.0	1.2	1.5	1.8	2.2	2.7
3.3	3.9	4.7	5.6	6.8	8.2

The values are repeated in multiples of 10.

Design tip — Marking capacitors

Values are often marked on the case (see photo, p.41). There is no room for this on small capacitors so the value is coded. The code has two digits. The first two digits are the first two digits of the capacitance, in picofarads. The third digit is the number of zeroes following the two digits.

Example

The code '223' means '22' followed by three zeros. This gives 22 000 pF, which is equal to 22 nF.

Tolerance is coded by an extra letter, as for the resistor printed code (p. 37).

Design tip — Selecting capacitors

For **high capacitance** (1 μF or more), use aluminium electrolytic capacitors. These may have axial or radial lead wires. Axial leads are one at each end of the capacitor. This can be useful if you need to 'jump a gap' on the circuit board. More often, especially on PCBs, you will need radial leads (both leads at the same end). If space is short, use tantalum bead capacitors, but these are more expensive.

For **medium capacitance** (10 nF to 1μF) normally use polyester or ceramic capacitors (not suitable for audio circuits). For better temperature stability, use polycarbonate.

For **low capacitance** (under 10 nF) use polystyrene or ceramic capacitors.

Capacitors have a **working voltage**, usually marked on the case. The capacitor is destroyed if this is exceeded. Polyester, polystyrene, polycarbonate and ceramic capacitors usually have working voltages of 100 V or more, so you will usually have no problems. Electrolytics have lower working voltages. If it is important to reduce leakage, choose an electrolytic with higher working voltage than you need (say, 63 V). However, it will be more expensive and larger than one with a lower working voltage (10 V or 25 V).

Electromagnetism

It can be shown that, when a current flows in a wire, a magnetic field is produced around the wire. If the wire is formed into a coil, the magnetic field resembles that of a bar magnet.

The magnetic field is represented by **lines of force** that show the direction of the field in and around the coil.

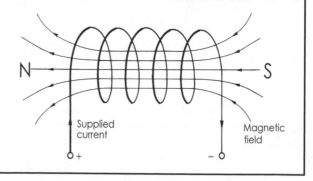

Induction

A current produces a magnetic field (see opposite) and the reverse action also occurs. In the drawing below, a bar magnet moving toward a coil **induces** a current in the coil:

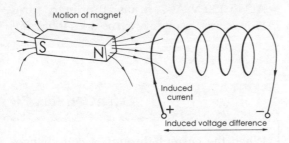

The current flows and the voltage difference is produced only when the magnet is *moving*. If we hold the magnet still, the current stops. If we move the magnet *away from* the coil, the current flows in the opposite direction.

Note that the current is flowing in the same direction in the drawing opposite and in the drawing above. This means that the induced current is producing a magnetic field with its north pole at the end nearer the magnet. This induced magnetic field is trying to *repel* the magnet (like poles repel). It is trying to stop it from moving toward the coil.

The induced magnetic field opposes the motion of the magnet. When we move the magnet away, the current reverses and the field reverses too. Now there is a south pole near the north pole of the magnet. Unlike poles attract. There is a force trying to prevent us from moving the magnet away.

Whichever way we move the magnet, there is a force to *oppose* the motion. We have to do extra muscular work to move the magnet. The extra energy we use appears as the voltage difference between the ends of the coiled wire.

Things to do

Connect a coil of wire to a multimeter switched to a low voltage range. Take a bar magnet and move it towards one end of the coil. Try it with the north pole nearer the coil, then with the south pole nearer the coil. Hold the magnet still. Move the magnet away from the coil. Move the magnet slowly. Move the magnet quickly.
What do you notice about the voltage?

Transformers

A transformer consists of two coils wound on a core. The core is made of layers of iron.

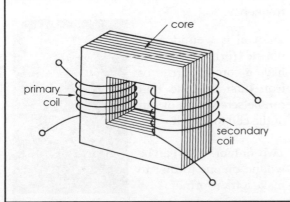

The coils normally have many more turns than are shown in the drawing.

When a current is passed through the primary coil it produces a magnetic field. The core is to provide a path for the lines of magnetic force so that they almost all pass through the secondary coil. Induction occurs only when there is a *change* of magnetic field. So a transformer does not work with DC. When AC flows through the primary coil there is an alternating magnetic field. This induces an alternating current in the secondary coil.

Transformer rules   Box 25

Frequency: The frequency of the induced AC equals that of the inducing AC.

Amplitude: If V_P is the amplitude of the voltage in the primary coil, and V_S is the amplitude of the voltage in the secondary coil, then:

$$\frac{V_S}{V_P} = \frac{\text{secondary turns}}{\text{primary turns}}$$

Example

A transformer has 50 turns in the primary coil and 200 turns in the secondary coil. The amplitude of the primary AC is 9 V. What is the amplitude of the secondary AC?

Rearranging the equation above gives:

$$V_S = V_P \times \frac{\text{secondary turns}}{\text{primary turns}}$$

$$V_s = 9 \times \tfrac{200}{50} = 36 \text{ V}$$

The amplitude of the secondary current is 36 V.

These calculations assume that the transformer is 100% efficient.

Self test

1 A transformer has 25 primary turns and 1200 secondary turns. The amplitude of the primary AC is 5 V. What is the amplitude of the secondary AC?

2 A transformer is used to transform 20 V AC to 120 V AC. It has 50 primary turns. How many secondary turns does it have?

Self induction Extension Box 26

When the current through a coil changes, the magnetic field through the coil changes. This changing field acts just like a magnet being moved around near the coil — it induces another current in the coil. The direction of the current opposes the change in current through the coil. This effect, in which a coil induces current *in itself*, is called **self induction**.

Self induction is important in the action of chokes (see below). If there is a very rapid change of current, such as when the current is switched off, self induction produces a very large current which may damage components in a circuit (p. 82).

Types of inductor   Extension Box 27

Chokes: They are used to block high-frequency signals (AC) from passing through from one part of a circuit to another. Low-frequency signals or DC voltage levels are able to pass through. Large chokes look like transformers, but have only one coil. Small chokes consist of beads or collars made of ferrite, threaded on to the wire that is carrying the high-frequency signals. Ferrite is an iron-containing material, so it acts as a core to contain the lines of magnetic force around the wire. Sometimes a choke is made by winding the wire around a ferrite ring.

Tuning coils: Used in radio transmittters and receivers to tune a circuit to a particular radio frequency.

The coil is wound on a plastic former. It may have a core of ferrite or iron dust ceramic that can be screwed in or out of the coil to tune it.

Two or more coils can be wound on one former to make a transformer.

When a wire is carrying a **current**, and that wire is in a magnetic **field**, there is a force that makes the wire **move**.

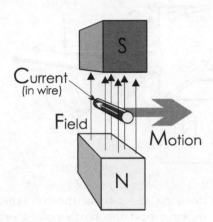

The direction of motion can be found by holding your left hand so that your thumb, your first finger and your second finger are pointing at right-angles to each other. Then:

- Point your **F**irst finger in the direction of the **F**ield (which goes North to South).

- Point your se**C**ond finger in the direction of the **C**urrent.

- Your thu**M**b is now pointing in the direction of the **M**otion.

Remember the three directions — **F**, **C**, and **M**! This is known as **Fleming's Left-hand Rule**.

How a simple DC motor works

The drawing at top right represents a simple DC motor. There is a permanent magnet to provide the field. A coil mounted on an axle spins between the poles. The coil is shown with a single turn. In a real motor, it would have several hundred turns.

The ends of the coil are connected to a **commutator**, consisting of two metal half-rings. There are two springy **brushes** in contact with the half-rings.

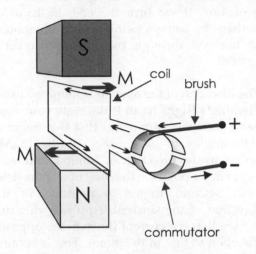

When a DC voltage is applied to the terminals, current flows along the upper brush to the commutator, around the coil to the other half-ring of the commutator and back along the lower brush. The current is flowing *away* from the commutator in the upper section of the coil. The field and current are as in the diagram on the left. Applying Fleming's Left-hand Rule, the upper section of the coil is forced to move to the right. Applying the rule to the lower section, where the current is flowing *toward* the commutator, that section is forced to move to the left. The two forces cause the coil to rotate in an anti-clockwise direction.

The coil spins round until the gaps between the half-circles come under the brushes. For an instant, the current stops. Inertia makes the coil continue to spin until the brushes contact the half circles again. But the section that is now the upper one still carries current away from the commutator. The section that is now the lower one still carries current toward the commutator. So the coil continues to turn in a clockwise direction.

In a real motor, the coil has a core of layers of iron (an **armature**) to help guide the lines of magnetic force through the centre of the coil.

Generators

A DC motor (p. 47) can also be used as a generator. If we turn the coil, by hand or perhaps by using a petrol engine, the motion of the coil through the field generates a current.

The direction of the current is found by using **Fleming's Right-hand Rule**. Hold your right hand with the thumb and first two finger at right-angles to each other. If the thuMb points in the direction of Motion, and your First Finger points in the direction of the field, your seCond finger is pointing in the direction of the Current. Applying this rule we see that the current flows in the opposite direction to that in the motor. This is because the induced current is trying to prevent the coil from rotating. To turn the coil we have to supply extra energy, which appears as generated electrical power.

An AC generator is similar, but has two slip rings instead of a commutator. These are connected to the ends of the coil. As the coil is rotated, each brush remains in contact wth the same ring.

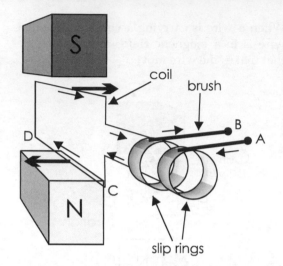

In the drawing, the part of the coil labelled CD is moving past the north pole. Current flows in through terminal A (negative) and out through terminal B (positive). As the coil is rotated, CD moves upward and then moves past the south pole. The current in it reverses. Now terminal A is positive and terminal B is negative. The current is alternating. It goes through one cycle for each rotation of the coil.

Power and current

Power is defined by the equation:

$$\text{power} = \text{current} \times \text{voltage}$$
$$P = IV$$

From Ohm's Law we know that:

$$V = IR$$

Putting this expression for V into the power equation, we get:

$$P = IV = I \times IR = I^2R$$

$$\mathbf{P = I^2R}$$

Example 1

A current of 5 A passes through a 16 Ω resistance. The power is:

$$P = 5^2 \times 16 = 25 \times 16 = 400 \text{ W}$$

Example 2

The current in the previous example is doubled to 10 A. The power now is:

$$P = 10^2 \times 16 = 100 \times 16 = 1600 \text{ W} = 1.6 \text{ kW}$$

Self test

1 The resistance of the coil of a DC motor is 10 Ω. The motor is being driven by 600 mA. What is the power?

2 A soldering iron heater coil has a resistance of 3.7 kΩ, and runs at 15 W. What current does it take?

Power and voltage

Another form of the Ohm's Law equation is:

$$I = \frac{V}{R}$$

Putting this expression for I into the power equation, we get:

$$P = IV = \frac{V}{R} \times V = \frac{V^2}{R}$$

$$\mathbf{P = V^2/R}$$

Example 1

A 6 V battery is connected to a 100 Ω resistor. The power is:

$P = 6^2/100 = 36/100 = 0.36$ W

Example 2

What is the maximum voltage that can be applied across a 220 Ω, 0.25 W resistor without damaging it? Given R and P, we calculate:

$$V = \sqrt{PR} = \sqrt{0.25 \times 220} = \sqrt{55} = 7.4 \text{ V}$$

Self test

1 A 2.2 Ω resistor has a 12 V battery connected across it. What is the power?

2 A 2 kW room heater runs on the 230 V mains. What is its resistance?

Mains power distribution

Mains power is generated in power stations, as outlined in Topic 6.

It is distributed to homes, offices, shops, factories and other places in which it is used by the National Grid. The National Grid is a network of cables spreading to all parts of the country. Power stations supply electricity to the Grid, the amount they supply depending on local demand at the time.

The cables of the Grid have thick copper or aluminium conductors, so their resistance is low. However the cables are many kilometres long and their resistance can not be ignored. As electricity passes along the cables there is a loss of power. Electrical energy is converted to heat energy, which escapes into the surroundings.

The power loss is proportional to the square of the current (see opposite). To keep the loss as small as possible, the current must be as small as possible. For a given power level, to make current as small as possible, the voltage must be as large as possible.

Power lines cross the Western Australian bush

Power is distributed through high-voltage power lines. These may run underground or may be above ground and supported by pylons (photo above). Underground power lines are necessary in built-up areas, but are more expensive to install. There are also complications in routing the power lines through areas where water mains, gas mains and sewers are already in position. It is relatively cheaper to string cables from pylons, which is a very important factor for lines many kilometres long in country areas. Unfortunately, most people consider rows of pylons to be unsightly.

High voltage transmission

Power is generated at about 10 kV, though this depends on the way in which it is generated. Before distributing it, a transformer at the power station steps up the voltage to 66 kV, 132 kV, or even to 400 kV. It is distributed through the National Grid at this voltage.

Before being supplied to the user, local transformers step it down (in stages) to 230 V.

The final stage of transforming to 230 V for use in a group of premises may be done by a transformer mounted on the power poles.

Transformers play an essential part in power distribution. Transformers work only with AC. This is one reason why mains power is supplied as alternating current.

Power losses during transmission can be estimated like this. Suppose that the resistance of the power line between the generating station and a house is 1 Ω. Suppose that the various appliances in the house are using power at the rate of 6 kW. Assume that the current is being distributed at 230 V.

First calculate the current:

$I = P/V = 6000/230 = 26$ A

Next calculate the power loss in the line:

$P = I^2R = 26^2 \times 1 = 676$ W

The loss is over 10% of the power being consumed in the house.

Now see what happens if the transmission voltage is raised to 132 kV. We can ignore the loss in the relatively short distance after the voltage has been transformed down to 230 V.

At the higher voltage, the current is:

$I = V/R = 6000/132\,000 = 0.045$ A

The power loss in the line is:

$P = I^2R = 0.045^2 \times 1 = 0.002$ W

The loss is only 0.000 003%. This is why electric power is distributed at high voltage.

Questions on resistance and capacitance

1 List four types of resistor and say when they are most likely to be used.

2 State the resistances of resistors that have bands of these colours:

(a) green, blue, brown, with a red tolerance band.

(b) brown, black, blue, with no tolerance band.

(c) grey, red, orange, with a gold tolerance band.

(d) orange, orange, yellow, with a silver tolerance band.

3 What colours are the bands on resistors of these values: **(a)** 100 kΩ ±5%, **(b)** 27 Ω ±20% , **(c)** 1.2 MΩ ± 10%?

4 A resistor is marked with four bands: yellow, orange, red, gold. What can you say about its actual resistance?

5 State the values and tolerances of capacitors marked as follows: **(a)** 473J, **(b)** 394K, **(c)** 102J.

6 What is the effective resistance of 680 Ω, 1.2 kΩ, and 56 Ω, connected in series?

7 What is the effective capacitance of 470 nF, 150 nF and 1.2 μF connected in parallel?

8 What are the main features of an aluminium electrolytic capacitor?

9 The range selector of the multimeter in the photo is set to 'Ohms × 10'. What resistance reading is shown?

10 Design a potential divider to produce a 4.7 V output from an 8 V input. The divider must be able to supply up to 10 mA without serious fall in output voltage.

11 Four wires join at one point in a circuit. The currents in three of them are: 34 mA toward the point, 62 mA away from the point, and 120 mA toward the point. What is the current in the fourth wire?

12 Three resistors, 120 Ω, 180 Ω, and 82 Ω, are joined in series with a 6 V battery. What is the current in the circuit? What is the voltage across each resistor?

Extension questions

13 What is the effective resistance of this network?

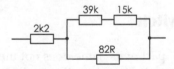

14 A bar magnet has its south pole nearer to one end of a coil. The bar magnet is being moved away from the coil. Draw a diagram to illustrate this and show the direction of the current induced in the coil.

15 Describe one type of choke and what it does.

16 A transformer has 24 V AC applied to its primary coil. The primary coil has 500 turns and the secondary coil has 40 turns. What is the voltage across the secondary coil?

17 A transformer has 6 V AC supplied to its primary coil and is to deliver 24 V AC from its secondary coil. The primary coil has 400 turns. How many turns has the secondary coil? If the frequency of the 6 V supply is 100 Hz, what is the frequency of the 24 V output from the transformer?

18 State Fleming's Left-hand Rule.

19 In the drawing of a DC motor below. State which terminal, A or B, is made positive in order to make the coil rotate in the direction shown. Name the parts labelled C and D.

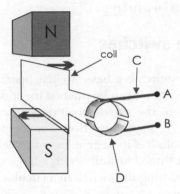

20 What is the name of the rule that relates the directions of motion, current and field in a generator?

21 Why does an AC generator produce an alternating voltage. If the coil is turned at 120 revolutions a minute, what is the frequency of the AC produced?

22 If a 43 Ω resistor is rated at 0.5 W, what is the maximum voltage that can be placed across it?

23 A pocket calculator uses power at an average rate of 8 mW. It is powered by two lithium cells delivering a total of 3 V. What is its effective resistance?

24 A current of 3 A flows through a low-voltage projector lamp when it is running at 35 W. What is the resistance of its filament when the lamp is running at this rate?

25 Why is mains power distributed at high voltage?

26 Why must mains power be distributed as an alternating current?

15 Switches

Switches are used to control the flow of current in a circuit. Current flows when the switch contacts come together. We say that the switch is **closed**, or contact is **made**. Current can not pass through the switch when the contacts are apart. We say that the switch is **open**, or contact is **broken**.

There are many different types of switch, used for different purposes. Here we describe a few of the most useful types.

Toggle switches

A toggle switch is a basic switch, operated by a toggle lever that can be pushed up or down. By convention, the down position is the 'on', or 'closed' , or 'made' position. The toggle switch in the photo has its toggle lever up. Behind the lever is a threaded dolly with a large nut. This is for mounting the switch in a circular hole cut in a panel.

At the rear are two terminal tags to which the connecting wires are soldered.

This switch is a heavy-duty type, rated to switch currents as large as 10 A and 250 V AC. Heavy-duty toggle switches are often used for switching the mains power supply to appliances and equipment. However, they can be used for switching smaller currents too.

This miniature toggle switch is suitable for mounting on a control panel.

It is able to switch 1.5 A at 250 V AC.

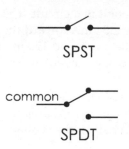

The larger toggle switch has two solder tags, showing that it has **single-pole, single throw** contacts (**SPST**). Its symbol shows how it works. It switches a single circuit and it is either open or closed.

The miniature toggle switch has **single-pole, double-throw** contacts (**SPDT**). The centre tag is common and can be switched into contact with either of the other tags. Such contacts are called **changeover contacts.**

Microswitch

The 'micro' part of its name does not mean that the switch itself is necessarily small. It means that the operating button moves only a small distance.

The switch is very sensitive. A light pressure on the lever causes the switch to click over from one position to the other. Most microswitches have SPDT contacts so that they can either switch something on or switch something off, perhaps both at the same time.

The contacts on a microswitch are sprung so that normally the common contact connects to what is called the **normally closed (n.c.)** contact. The third contact is the **normally open (n.o.)** contact.

Microswitches are used where a switch has to be operated mechanically. For example, a microswitch is mounted inside a cupboard, so that the lever is held down when the door is closed. Its common and normally closed contacts are wired into a lamp circuit. When the door is closed, the contacts are open and the lamp is off. When the door is opened, the contacts close and the lamp is switched on.

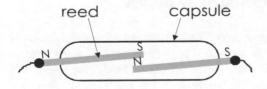

Reed switch

This consists of two springy contacts (the reeds) sealed into a capsule with their ends over-lapping. Normally the contacts are open. In a magnetic field, the reeds become magnetised. The north and south poles attract each other. The reeds come together and make contact. When the field is removed, their springiness moves the reeds apart.

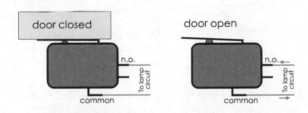

Larger reeds switches can switch mains currents of up to 2 A.

The magnetic field may be provided by a permanent magnet or by a coil. For example, in a security system, a reed switch is mounted on a door-frame. A magnet is mounted on the door. When the door is shut, the magnet is near to the switch. Contact is made. If the door is opened, the magnet is moved away from the switch. The contacts open, breaking the circuit and triggering an alarm.

If the reed switch is operated by a coil wound round its capsule, the switch acts as a relay (see p. 55).

Tilt switch

Two contacts are sealed in a glass capsule containing a large drop of mercury. When the capsule is upright the mercury settles at the bottom and bridges the gap between the contacts. The switch is closed.

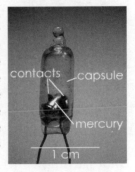

If the switch is tilted, the mercury runs to the side or the other end of the capsule. There is no longer a bridge between the contacts. The switch is open.

Mercury switches are often used in electrically powered machines. They are attached to a movable part of the machine and used to detect when that part is in the correct position. There is a risk of the capsule breaking and spilling mercury, which is harmful. Mercury free tilt switches are available.

Self test

What type of switch would you use for:

(a) detecting if a protective grid on a lathe is in position.

(b) a power switch on a PSU.

(c) detecting if a mobile robot has fallen over.

(d) the 'click' button on a computer mouse.

Push switches

Λ push switch is operated by pressing a button. There are two types of action. Most switches are **push-to-make** (or **PTM**) switches. Pressing the button pushes the contacts together and the switch closes. The other type are **push-to-break** (or **PTB**) switches. The contacts are normally closed but are forced apart when the button is pressed.

Either type of switch may be momentary or latching. A switch that is **momentary** acts only for as long as the button is pressed. When the button is released, the switch springs back to its normal state.

In a **latching** switch, the button stays down when pressed. The contacts remain closed or open, depending on the type of switch. You need to press the button *again* to return the button to its normal state.

Push buttons are used for a wide variety of control purposes, and may be used for power switching for lamps, radio sets and other appliances.

Rocker switches

These are similar in action to toggle switches and are used for the same purposes. The difference is that they are operated by a rocker.

Slide switches

These have uses similar to toggle switches, but they are operated by a sliding knob.

The slide switch in the photo is a SPDT switch. The common connection is made to the central terminal.

DPDT switches

Toggle, rocker and slide switches are also made as double-pole double throw versions. This gives two separate switches within the same unit, but operated together.

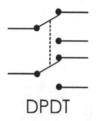

DPDT

DPDT switches can be used to switch on two circuits at the same time. They may also be wired into both the live and the neutral mains lines.

When the switch is off, the appliance is completely isolated from the mains. DPDT switches are often used with electric blankets for safety.

Key switches

A key switch can be turned on and off only by using a key. Only the pair of keys sold with the lock will allow the switch to be operated. Key switches are used when security is important.

Keyboard switches

These are PTM, momentary switches with large square tops, often marked with letters, number or symbols. They are used for building keyboards for computers. Cheaper keyboards use **membrane switches,** in which a conductive plastic sheet is pressed against contacts beneath.

Rotary switches

Rotary switches are used for switching one line to any one of several other lines. Often several such switches are combined in one unit.

They are used for functions such as selecting wavebands on a radio receiver, or selecting the measurement ranges of a multimeter.

The switch has one or more rotating contacts surrounded by a ring of usually 12 stationary contacts. Switches are produced with several different arrangements of contacts. These are 1-pole-12-way, 2-pole-6-way (see drawing), 3-pole-4-way and 4-pole-3-way.

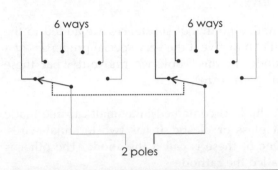

Relays

A relay is a current-controlled switch. Compare with a reed switch, p. 53

A relay has a low-voltage coil wound on a core. There is an iron armature that is attracted toward the core when current passes through the coil. This is attached to a sprung lever. The common contact moves across from the normally closed contact to the normally open contact.

Most modern relays are totally enclosed and sealed in, like the one below. Most have SPDT contacts, but there are also DPDT versions. The larger ones will switch 10 A at 250 V AC. The maximum voltage for switching DC is always much less, often only a half of the maximum AC voltage. There are also miniature relays like the one below that are suitable for mounting on circuit boards.

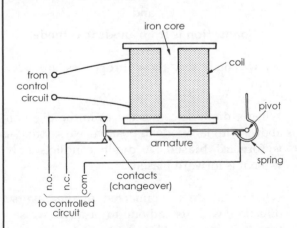

A typical relay of this types operates in about 10 ms.

These are the symbols used in circuit diagrams for the relay coil (left) and the changeover contacts (right). The solid arrow indicates the normally closed contact.

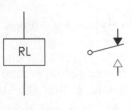

16 Diodes

A diode is made of silicon. Silicon is neither an insulator nor a conductor. It is a **semiconductor**. This means that its properties are different from ordinary conductors such as copper.

Small amounts of substances are added to the silicon to give it the very special properties of a diode. In this Topic we find out what these properties are.

A diode is contained in a small capsule made of glass or plastic. It has two terminal wires. One of these is called the **anode**. The other is called the **cathode**.

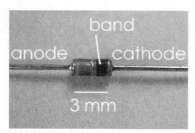

Usually there is a band marked on the diode to show which wire is the cathode.

Things to do

You need:
- A 6 V battery or a PSU.
- A 6 V lamp in a socket, with connecting wires.
- A diode. Type 1N4148 is suitable, but any ordinary silicon diode will do.
- A breadboard.

1 Examine the diode. Look for the band and identify the cathode wire.
2 Set up the circuit shown at top right on this page, but not the battery or PSU.
3 Check that the diode is connected the right way round. Its cathode wire goes to one terminal of the lamp.

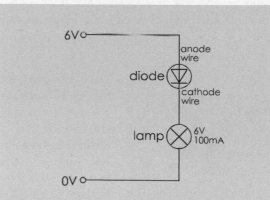

4 Connect the battery and see what happens to the lamp. What can you say about the diode?

5 Disconnect the battery. Remove the diode and replace it the other way round. It now has its anode wire connected to the lamp.

6 Connect the battery and see what happens to the lamp. What can you say about the diode now?

Conduction through a diode

The experiment above demonstrates that:

> **A diode conducts in only one direction**
>
> and
>
> **Conduction is from anode to cathode.**

As we shall see, these properties are very useful.

When a diode is connected as in the diagram above, with its anode to positive, we say that it is **forward biased**. A diode conducts only when it is forward biased.

When a diode is connected in the reverse direction, with its cathode to positive, we say that it is **reverse biased**. A diode does not conduct when it is reverse biased.

Voltage drop

In a voltage divider circuit (p. 35), the output voltage is a proportion of the input voltage. The values of the two resistors decide what the proportion is. We will see what happens if we replace one of the resistors with a diode. The diode is forward biassed so that current can flow though it.

Things to do

You need:
- 6 V battery or PSU.
- 220 Ω resistor.
- diode.
- multimeter or 2 V voltmeter.
- breadboard.

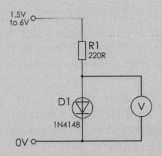

1 Set the input voltage to 6 V. Measure the voltage across the diode. Record the result in a table.

Input voltage	Output voltage
6	
4.5	
3	
1.5	

2 Repeat (1) with the voltage set to 4.5 V, 3 V, and 1.5 V.

3 What do you notice about the output voltage as you change the input voltage?

4 Is output proportional to input?

5 Does this circuit work like a voltage divider?

The results of the investigation show that a diode does not behave like a resistor. It does not obey Ohm's Law. The output voltage (that is, the voltage across the diode) varies only slightly with input voltage. It stays very close to 0.7 V.

We can sum up the results of the investigation by saying that:

A forward biased diode has a voltage drop of about 0.7 V.

This voltage drop is called the **forward voltage drop**.

Testing diodes

Many multimeters have a **diode test** function. It measures the forward voltage drop of the diode.

Things to do

You need:
- several diodes of different types, including some faulty ones.
- multimeter that has the diode test function.

1 Turn the range selector knob to the diode test function.

2 Connect the test probes of the meter to the diode, with the negative (black) probe to the cathode and the positive (red) probe to the anode.

3 With a good diode the reading is about 0.7 V (700 mV). Diodes vary slightly, so it may range between 400 mV and 900 mV. Anything outside this range indicates a faulty diode.

4 Reverse the connections to the diode. There should be a 'zero' or equivalent reading (depending on the meter) to indicate that the diode is not conducting when reverse biased.

17 Rectifier diodes

One of the most important uses of diodes comes from their ability to conduct in only one direction. See what happens in this circuit:

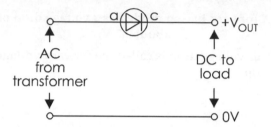

The supply is alternating current, from a transformer.

In the diagram below, there is a load connected to the circuit. The diagram shows the path of the current during the positive half-cycles of the AC. The diode is foward biased, so it conducts. Current flows through the diode to the load and returns along the 0 V line.

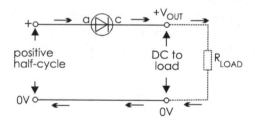

The diode does not conduct during the negative half-cycles, as shown below:

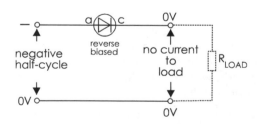

The waveform of the current through the load is plotted at top right. Although the voltage is pulsing, it is always positive. It is the equivalent of DC.

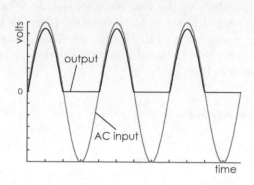

Comparing the graphs of the AC input and pulsed DC output, we see that:

- There is no output during negative half-cycles. Half the input power is wasted.

- The output amplitude is less than the input amplitude. This is because of the forward voltage drop across the diode.

A circuit that converts AC to DC is called a **rectifier**. Because this rectifier produces current only from the positive cycle, it is called a **half-wave rectifier**.

Full-wave rectifier

The circuit below rectifies AC by using a **bridge** of four diodes:

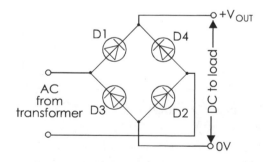

During the positive half-cycle, diodes D1 and D2 are forward biased, so they conduct. Diodes D3 and D4 are reverse biassed and do not conduct. Current flows through the load as shown in the next diagram.

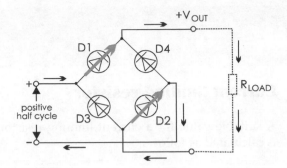

During the negative half-cycle, diodes D1 and D2 are reverse biased so they do not conduct. Diodes D3 and D4 are forward biased and conduct.

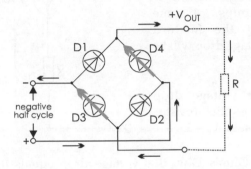

The result is that current continues to flow through the load, *in the same direction* as before. The graphs of input and output are:

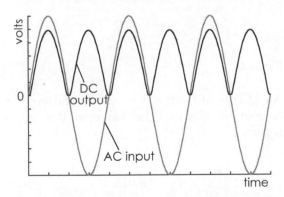

The rectifier produces output during both half-cycles, so it is 100 % efficient. It is called a **full-wave rectifier**. In each half-cycle, the current flows through *two* diodes, so the output amplitude is *two* voltage drops (about 1.4 V) less than the input amplitude.

Rectifiers are used in PSUs and other power units to produce DC from the low-voltage output of a mains transformer.

Smoothing

The pulsed DC from a rectifier is unsuitable for powering circuits until it has been smoothed. This is done by connecting a large-value capacitor across the DC output.

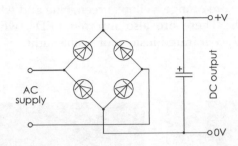

The capacitor is usually an aluminium electrolytic one and has a capacity of 1000 μF or more. Repeated pulses of DC soon charge the capacitor up to the peak voltage. When current is being drawn from the circuit the voltage starts to fall after each peak, but it is returned to peak level by the next pulse. The result is DC with a slight **ripple**.

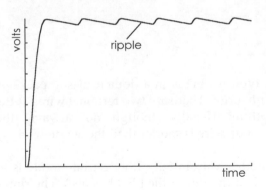

If the capacitor is large enough and the current drawn by the load is not too large, the voltage is almost as smooth as a pure DC.

Things to do

Use a low-voltage mains transformer and an oscilloscope to view the output from half-wave and full-wave rectifiers, unsmoothed and smoothed

18 Light emitting diodes

Light emitting diodes, usually known as **LEDs**, give off light when a current flows through them. The original LEDs were red, but now there are orange, yellow, green, blue and white LEDs. There are also infrared LEDs, which produce infrared instead of visible light.

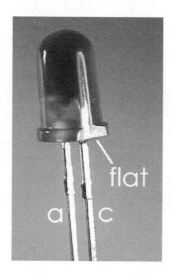

A typical LED is in a domed plastic package, with a rim. There are two terminal wires at the bottom. Usually, though not always, the cathode wire is shorter than the anode wire.

Another way to tell cathode from anode is to look at the rim (if the LED has one). The rim is flat on the side nearer the cathode wire.

An LED needs about 20 mA to light it to full brightness, though as little as 5 mA still produces a clearly seen glow. The forward voltage drop (p. 57) of an LED averages about 1.5 V, so a 2 V supply will light most types to their maximum brightness. When lit by a higher voltage, the LED may be burnt out if the forward voltage across it exceeeds 2 V. It is essential to wire a **current limiting resistor** in series with it.

Current limiting resistor

A suitable value for a current limiting resistor is calculated as follows.

The supply voltage is V_S volts. The current that we want to flow through the LED is i amps. Assume that the forward voltage drop will be 2 V.

The voltage drop across the resistor must be $V_S - 2$.

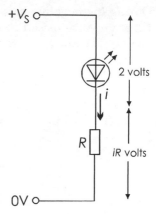

By Ohm's Law, this voltage drop equals iR. Therefore:

$$V_S - 2 = iR$$

Rearranging terms gives:

$$R = \frac{V_S - 2}{i}$$

Example

An LED is lit from a 9 V supply and takes a current of 15 mA. The value of the series resistor is:

$$R = \frac{V_S - 2}{i} = \frac{9 - 2}{0.015} = 466 \, \Omega$$

Use the next higher resistance in the E24 series, which is 470 Ω.

Self test

Calculate the value of the series resistor required for an LED that:
(a) runs at 25 mA on a 12 V supply.
(b) runs at 10 mA on a 15 V supply.

Shapes and sizes

LEDs are used as indicator lamps. For example, to indicate that the power to an appliance is switched on. They are also used for informative or decorative displays. They are made in a variety of shapes, including circular, rectangular, and triangular LEDs

Arrays of shaped LEDs are used for displays. The commonest of these is the seven-segment display, used for displaying numerals and letters.

One or more rows of these are used for displaying messages.

LEDs are made in a range of sizes. The smallest are about 1 mm in diameter, used as indicators on panels where space is short. At the other extreme the jumbo LEDs, 10 mm in diameter, are useful where a readily noticeable warning lamp is needed.

LEDs are ideal as indicators because they require very small currents when compared with filament lamps. This makes them very suitable in battery-powered equipment, in which a filament lamp would soon exhaust the supply. There is also the factor that filament lamps have a limited life. Sooner or later the filament burns out. LEDs last almost for ever.

Reverse bias

An LED is not able to withstand a reverse bias of more than a few volts. Most can survive a reverse bias of 5 V, but no more. This is quite different from most diodes, which can withstand reverse bias of hundreds of volts.

Since the circuits in which LEDs are used often have a supply voltage of 6V or more, it is important to make sure that the diode is wired into the circuit the right way round.

Bicolour LEDs

An LED that can change colour is useful in some applications. For instance the LED could indicate 'all systems GO' when it is green and 'fault condition' when it is red. On a digital camera, there may be a 'record/play' indicator that is red when the camera is in recording mode and changes to green when it is in play mode. LEDs that can display two colours are called **bicolour LEDs**.

A bicolour LED has two LEDs of different colours inside one package. There are two ways in which the LEDs may be connected.

In the type with three terminal wires, the LEDs have a common cathode. A positive voltage applied to either of the other two wires is used to light the corresponding LED.

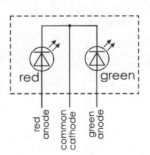

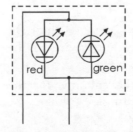

In the two-terminal type the LEDs are connected anode to cathode. Which LED lights depends on which terminal is made positive.

By switching the supply, we can light red and green alternately. If this is done at high speed, the LED appears to be emitting yellow light.

Power diodes

The diode illustrated in the photo on p. 56 is a signal diode. It is able to pass a maximum current of 100 mA. A rectifying circuit for a power supply unit generally needs to pass more current than that. Special power rectifying diodes are made that can pass much larger currents.

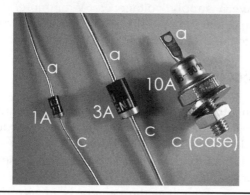

Below is a **bridge rectifier**. This consists of four power diodes in one package. They are already connected to make a full-wave rectifying bridge. It has four terminal wires, two for the AC input and two for the DC output.

Voltage/current graphs

The relationship between the voltage across a diode and the current through it is investigated below. First we test the diode when it is connected with forward bias.

Things to do

Set up this circuit on a breadboard:

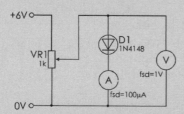

Check that the diode is connected the right way round, with its cathode wire to the ammeter.

1 Draw a table of three columns for the results. Head the columns 'Voltage, V', 'Current, I' and 'Ratio V/I'.

2 Connect the power supply and turn VR1 so that the voltmeter reads zero. Record voltage and the current in the table.

3 Turn the knob of VR1 slowly until the voltage across the diode is 0.1 V. Record this value. Read and record the current.

4 Repeat step (3) for increasing voltages: 0.2, 0.3, 0.4, 0.5, 0.6, 0.7, 0.9, 1.0, 1.5 and 2 V. You may need to change the ammeter or the range of the multimeter to an fsd of 1 mA for later readings. Note: 'fsd' means 'full scale deflection', the maximum reading.

5 Calculate the ratio V/I for each pair of readings. Does this remain constant for all pairs? Compare this result with the result from p. 28.

6 Plot a graph of your readings of voltage and current. In your own words, describe how the current changes as voltage is increased.

7 Repeat steps 1 to 6 with the diode connected the other way round, so that it is reverse biased. Range the voltage up from 0 V to 10 V in steps of 1 V. Record your results.

8 How much current flows through a diode when it is reverse biased?

Current through a diode

The investigation opposite tells us several things about the current through a diode:

When forward biased:

- No current flows when the voltage is less than about 0.6 V.
- When the voltage is a little over 0.6 V, a small current flows. With higher voltage, a large current flows.

When reverse biased:

- No current flows.

No diode is perfect, so when we said (above) that there is no current through the diode, there is always a small leakage current. This is only a few nanoamps, which is too small to measure easily.

Zener diodes

If a Zener diode is reverse biased with a small voltage, it behaves like an ordinary diode. It does not conduct.

If a Zener diode is reverse biased with a voltage greater than a certain amount, called the **Zener voltage**, it conducts easily.

The Zener voltage of a Zener diode is fixed when the diode is manufactured. It may range from about 2.7 V to 20 V, with a tolerance of ±5%.

Zener diodes are used to regulate the output of power supply circuits. The circuit on the right has a rectifier with smoothed DC output (p. 59), followed by a Zener diode voltage stabiliser. The Zener diode is chosen with a Zener voltage equal to the required output voltage.

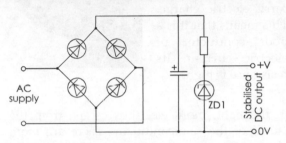

The output from the rectifier is several volts greater. Part of that voltage is dropped across the resistor. Its resistance is such that, when the load is drawing its maximum current, about 5 mA flows through the Zener to the 0V line. The output voltage is equal to the Zener voltage. If the load is taking less than its maximum, or no current at all, the surplus current flows away through the Zener diode to ground. The Zener diode is still operating and the output voltage is still equal to the Zener voltage.

Questions on diodes

1 What are the properties of a diode?

2 The cathode of a diode is made 4 V positive of its anode. What is the name for this type of bias? Does current flow though the diodes?

3 Draw a circuit diagram of a half-wave rectifier. Describe how it works. Draw a sketch of 3 cycles of its output waveform.

4 Answer question 3 for a full-wave rectifier.

5 How do we smooth the output of a rectifier?

6 What are the properties of an LED? Describe some of the ways in which LEDs are used.

7 With the help of a diagram, describe one type of bicolour LED and how it works.

Extension questions

8 Describe how the current through a forward biassed diode varies as the voltage is increased from 0 V to 3 V.

9 State the properties of a Zener diode. How is a Zener diode used as a voltage stabiliser?

19 Light dependent resistors

A light dependent resistor (or **LDR**) consists of a disc of semiconductor material with two electrodes on its surface.

In the dark or in dim light, the material of the disc has a relatively small number of free electrons in it. There are few electrons to carry electric charge. This means that it is a poor conductor of electric current. Its resistance is high.

1 cm

In the light, more electrons escape from the atoms of the semiconductor. There are more electrons to carry electric charge. It becomes a good conductor. Its resistance is low. The more light, the more electrons, the lower the resistance.

Things to do

This investigation is to measure the changes in the resistance of an LDR when we change the amount of light falling on it.

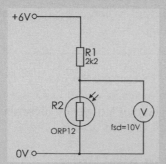

It is better to use a battery for the power supply, so that the circuit can be carried about.

1 The circuit is on the workbench to begin with. Connect the power supply and measure the voltage. Record your result.

2 Cover the LDR with your hand. Repeat the reading .

3 Move the circuit to other places and measure the voltage when it is in each place. Suitable places are: by a large window, outdoors, under a bright bench-light, under the workbench, in a cupboard.

4 In what kind of place is the voltage reading greatest? In a bright place or in a dark place?

5 What is the effect of light on the resistance of the LDR?

6 What kind of network is made by R1 and the LDR? (*Hint*: see p. 35)

7 What result would you get if you exchanged R1 and the LDR? Try it.

8 How could you alter the circuit to make it more sensitive to changes in light when it is in shaded places?

Circuits and systems

Leaving out the precise details of how the components are connected and how they work, we can represent the circuit on the left by a **system diagram**:

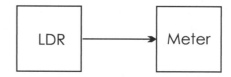

This shows that the system has two parts:
- The **LDR**, which receives **input** from the surroundings: an amount of light.
- The **meter**, which provides **output,** in this case, a measurement of voltage.

In this system, the two parts are directly linked. As we shall see later, most systems have more than two parts.

Drawing a system diagram makes it easier to understand what a circuit does. A system diagram is particularly useful when the circuit is complicated, with many parts.

Design time

Design time pages are scattered throughout the book. They provide collections of simple circuits, tips, problems, data and other things of interest to people designing circuits.

You are not expected to memorise these circuits. They are here to give you something to think about. The thinking will help you understand electronics better. You may find one or two of the circuits help you with your exam project. Or you might just like trying them out on a breadboard to see what happens.

The circuits are not described in detail. You are expected to work from the circuit diagram and think things out for yourself.

An LDR controls an LED. What happens? How does it work?

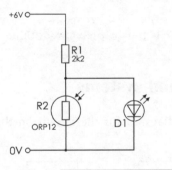

The LED in the circuit could be at the end of a long lead. It could be a **remote indicator**. This circuit could be improved (and will be later), but — for the moment — suggest some ways that this circuit could be used.

Try exchanging R1 and R2.

Shade the circuit, if necessary, so that the LED *just* goes out. Now change R1 for another resistor so that the LED comes on again. Try to find a resistor so that the LED is off in fairly dim light and comes on in very dim light.

What is the purpose of this arrangement of switches?

Add a switch to the circuit so that you can switch it to operate in (1) bright light, or (2) dim light.

When a DPDT switch is wired like this, what does it do?

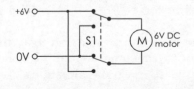

20 Thermistors

A thermistor is made from a semiconductor material. It is shaped into a disc, a rod or a bead. Bead thermistors may be only a few millimetres in diameter. Some bead thermistors have the bead enclosed in a glass capsule.

Because of their small size, bead thermistors respond very rapidly to changes of temperature. Thermistors have two terminal wires. The resistance of most thermistors decreases as temperature increases. These are **negative temperature coefficient** thermistors, or **ntc** thermistors. Thermistors with a positive temperature coefficient (ptc) are also made, but are less often used.

Thermistors are used in circuits that measure temperature or respond in other ways to temperature. They may also be used in circuits that could be put out of adjustment by changes of temperature. The thermistors automatically compensate for the temperature changes.

Things to do

This is a circuit for investigating the change of resistance of a thermistor with temperature. Preferably use a bead thermistor.

The thermistor is soldered to a pair of leads about 20 cm long, so that it can more easily be placed in various situations.

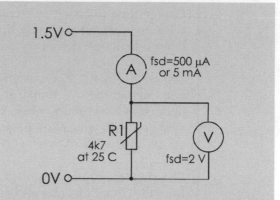

1 Put the thermistor in different places, with a thermometer beside it. Leave them there for 2 minutes. Then measure the temperature, current and voltage.

2 Record the results in a table that has a fourth column for the calculated resistance of the thermistor (V/I).

3 Plot a graph of resistance against temperature. Is the graph a straight line?

Circuits and systems

The system diagram for the circuit on the left is:

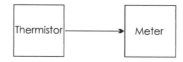

This has the same shape as the system diagram for the light meter (p. 64). Both diagrams can be drawn like this:

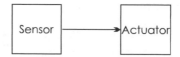

A **sensor** senses what is happening. Examples of sensors are switches, LDRs, and thermistors. An **actuator** makes something happen. Examples of actuators are motors, meters, lamps and LEDs.

Design time

A voltage divider circuit is the basis of this simple electronic thermometer. If you want to measure temperatures around room temperature (25°C), R2 should have about the same resistance as the thermistor R1 has at that temperature.

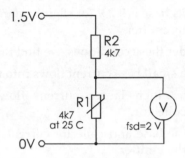

1.5V

R2
4k7

R1
4k7
at 25 C

V
fsd=2 V

0V

1 Stick an adhesive label on the outside of the meter to cover its scale. Use a removable label that can be peeled off later.

2 The first step is to **calibrate** the electronic thermometer. Put the thermistor and an ordinary room thermometer in various places that are at different temperatures.

3 Leave them for a few minutes in each place. Then draw a pencil line on the label, level with the needle of the meter. Mark the line with the temperature shown on the room thermometer.

4 Using the pencilled marks as a guide, draw a scale in ink to show temperatures in steps of, say, 5 degrees. Rub out the pencilled marks and figures.

5 Put the thermistor in several other places and read the temperature directly from your scale on the meter.

The range of the electronic thermometer depends on the value of R2. Design a circiit that includes a switch, so that there are two temperature ranges. Calibrate each range separately.

Add an indicator LED to this circuit, to show when the power is switched on.

It is a good idea to use a low supply voltage for these thermistor circuits. Why?

All the windows and doors of a house have a magnetic reed switch mounted on them. The switches are closed when the windows and doors are closed. Design a circuit that lights an LED only when *all* of the windows and doors are closed.

EVEN IF you are not following a Design and Technology course, these Design Time projects and problems will help you understand electronic theory.

21 Transistors

There are several classes of transistor. The class that is described in this Topic is known as a **silicon npn transistor**. It is also known as a **bipolar junction transistor**, or **BJT**. We shall not go into the reasons for these names, and you need not remember them, except for 'BJT'.

The BJT class of transistor described in this Topic is one of the most widely used. Another commonly-used class is the MOSFET, described in Topic 24.

All transistors have three terminal wires or connections.

Low-power transistors are enclosed in a plastic package or a metal case. The plastic case has a flat surface and the metal case has a tag on its rim. These are to help you to identify the terminal wires.

Seen from **below**, the wires are arranged like this in most (but not all) low-power transistors:

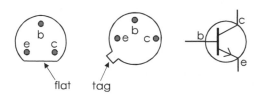

The symbol on the right is used to represent BJTs in circuit diagrams. The terminals are labelled with letters c, b and e, to identify the terminals as **collector**, **base** and **emitter**.

Transistor action

To use a BJT, we connect it so that:

- its **emitter** is its most negative terminal.
- its **collector** is several volts positive of its emitter.
- its **base** is 0.7 V (or slightly more) positive of its emitter.

Under these conditions, we find that:

- a small base current flows **into the base.**
- a much larger current flows **into the collector**.
- the base and collector currents flow **out of the emitter.**

This diagram illustrates the way the currents flow:

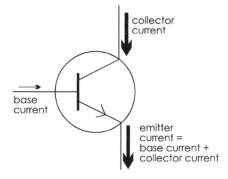

The base current is drawn with a thinner arrow because it is a much smaller current than the collector or emitter currents.

Things to do:

This is a demonstration of the relative sizes of the base and collector currents. We demonstrate the sizes of the current by passing them through filament lamps. The bigger the current, the brighter the lamp.

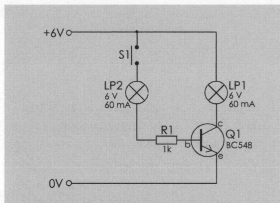

S1 is a push-to-make push-button. It has two short wires soldered to its terminals. Their other ends are stripped so that they can be pushed into sockets in the breadboard. LP1 and LP2 are two identical lamps, in sockets with connecting wires attached. Q1 is a transistor, type BC548; most other types can be used. A BC548 has the arrangement of terminal wires as in the diagram on p. 68.

1 Check that the three terminal wires of Q1 are inserted into the correct sockets.

2 Connect the power supply. Does LP1 light?

3 Press S1. Does current flow through LP1?

Does current flow through LP2?

4 Keep S1 pressed down and unscrew LP2 from its socket. Watch what happens to LP1. Was LP2 carrying a current at step (3)?

5 If so, what can you say about this current?

Transistor switch

The investigation above demonstrates one of the two important uses of transistors. A very small base current switches on a much larger collector current. We call this a **transistor switch.**

An an example, we can use the small current through an LDR sensor to switch a relatively large current through a filament lamp.

Here is the circuit:

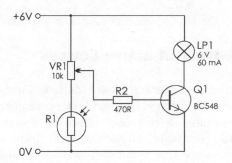

The LDR sensor consists of a voltage divider made up from VR1 and the LDR. A variable resistor is used so that the switching level can be set to different light levels. The transistor switch consists of R1 and Q1. R1 limits the amount of current being drawn from the potential divider. The collector current of Q1 flows through lamp LP1, and is about 60 mA.

With the LDR in normal room lighting, we adjust VR1 so that the light just goes out. When the LDR is shaded, its resistance increases. This increases the voltage across it.

An increase of voltage across the LDR raises the voltage at the wiper of VR1. More current flows to the base of Q1. More current flows through LP1 and into the collector of Q1. The lamp is switched on.

Things to do

Check that the switching circuit above works as described.

Circuits and systems

The switching circuit is a three-stage system:

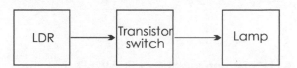

We will build several more systems that have this pattern.

22 Transistor action

Passive and active devices

Electronic components are of two kinds — **passive** and **active**. Passive devices are not able to generate an increase of power. Examples include resistors, capacitors and inductors. Resistors are able to convert electrical power to heat. Inductors are able to convert electrical energy into magnetic force. But neither of these devices are able to increase the power in the circuit. They are passive. In contrast, a transistor has a low-power input (small current) and converts this to a high-power output (large current). It is an active component. The energy for this activity comes from the electrical supply to the circuit.

Current and voltage changes

The investigation below looks in more detail at the changes of current and voltage in a transistor switch circuit.

Things to do

This circuit has two meters, a microammeter to measure the base current and a milliammeter to measure the collector current. You will also need a digital voltmeter (fsd=10 V) , but this is not to be connected into the circuit.

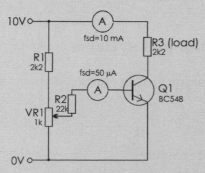

10V○

fsd=10 mA

R3 (load)
2k2

R1
2k2

fsd=50 μA

R2
22k

Q1
BC548

VR1
1k

0V○

The diagram specifies a BC548 BJT, but you can try it with other types, such as BC337, 2N2222A, or 2N3904.

1 Adjust VR1 so that there is zero base current (i_b). Read the collector current (i_c). Record your results in a table

2 Without altering the setting of VR1, use the voltmeter to measure v_{be}, the voltage difference between the base and emitter. Also measure v_{ce}, the voltage difference between collector and emitter.

3 Repeat steps (1) and (2) with i_b equal to 5 μA to 50 μA (if possible) in steps of 5 μA.

4 Plot a graph of collector current against base current. What does this tell you about the relationship between the currents?

5 Plot a graph of base voltage against base current. What does this show as base current increases?

6 Plot a graph of collector voltage against base current. What does this tell us about the voltage across the load as collector current increases?

Transistor action

The results you obtain from the investigation may vary slightly depending on the type of transistor tested. With a typical transistor, the graph of collector current against base current looks like this.

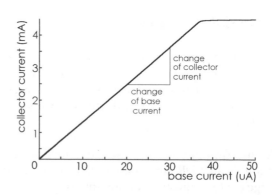

There is a straight-line (linear) relationship between the base and collector currents. In other words:

Collector current is directly proportional to base current.

We can find out something else of interest from this curve. We mark off a section of the curve and measure the *change* in base current. The section marked starts at 20 μA and runs to 30 μA. This is a change of 10 μA. Over the same section of the curve, the collector current changes from 2.5 mA to about 3.5 mA, a change of 1 mA (= 1000 μA).

Putting this all into microamps, we can say that a change of 10 μA in base current results in a change of collector current of 1000 μA. The change of collector current is 100 times bigger than the change of base current. Putting this in other words we can say that:

The current gain of the transistor is 100.

This current gain is usually called the **small signal current gain**, and it has the symbol h_{fe}.

The graph of base-emitter voltage against base current is like this:

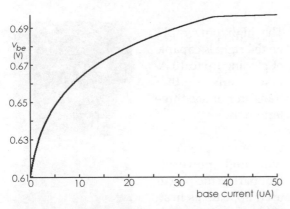

Noting the scale on the left, we see that the voltage between the base and emitter starts off a little less than 0.7 V and finishes up at 0.7 V. As an approximation we may say that:

The base-emitter voltage is close to 0.7 V.

In fact, it is equal to one diode drop, because there is the equivalent of a forward-based diode between the base and emitter terminals.

The collector-emitter voltage changes in this way:

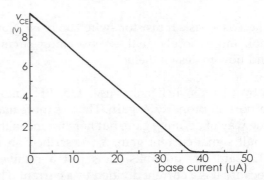

As base current increases, so does collector current. As the current through R3 increases, the voltage across it increases too (Ohm's Law). The voltage at one end of R3 is fixed at 10 V, the supply. The voltage at its other end (the collector end) must fall. The graph shows the falling voltage at the collector. It falls steadily until it is only slightly greater than zero. At this point it can not fall any further because the collector is only just positive of the emitter and the transistor would not work if it went down further. We say that the transistor is **saturated.** Or we can say that it has **bottomed out**.

In this circuit, the transistor saturates when the base current is about 37 μA. From the graph of collector current against base current, we can see that, when the base current reaches this value, the collector current levels off. It no longer increases in proportion to the base current.

Self test

1 What is the normal value of the base-emitter voltage?

2 What is the collector voltage of a saturated BJT?

3 In a given BJT, a change of 20 μA in the base current produces a change of 2.4 mA in its collector current. What is its small signal current gain?

23 Transistor switches

There is a basic transistor switch on p. 69. Now look more closely at these switching circuits and how to design them.

A transistor switch makes use of the BJT's most important property — **gain**. There is more than one way of defining gain, but here we mean the small signal current gain, h_{fe}, described on p. 71. Gain has no units. It is just a number, because it is a current divided by a current. The gain of a typical BJT is 100.

This circuit is an interesting but simple way of demonstrating transistor gain:

Things to do

The parallel lines A and B in the diagram represent two metal contacts spaced with about 1 mm between them. They could be two drawing pins pushed into a small scrap of wood, cork or plastic. Or they could be a pair of stripped copper wires, about 3 cm long, pushed into the sockets of a breadboard, and lying flat on the board. Select the sockets so that the wires have NO contact with each other, either electrical or physical.

Currents are small in this circuit, so there is no need for a resistor in series with the LED, or for one in the connection to the base of Q1.

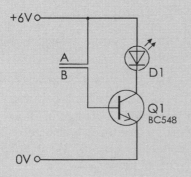

1 Check that there is no connection between A and B.
2 Connect the power supply. The LED should not light.
3 Press your finger on A and B to bridge the gap between them (but NOT to force them into contact with each other!)
4 The LED should light. You may not see it easily if you are close to a window. Try moistening your finger-tip before touching the contacts.
5 Estimate how much current is passing through the surface of your finger-tip.

Power transistors

Low power BJTs such as the BC548 are suitable for switching on LEDs and small filament lamps. They are rated to carry a collector current of up to 100 mA. Many other devices such as DC motors and bright lamps take more current than this. To switch these, we need medium-power or high-power transistors.

The high power BJT on the right is capable of passing up to 10 A. Like any other transistor, it has three terminals.

One of the problems with high power is that part of the power used appears as heat. With a current of several amps, the heat may be so great that the BJT is damaged.

To avoid this, we bolt a **heat sink** to the tag at the top of the transistor. The heat sink carries the heat away to the surroundings.

A heat sink is made of metal (usually aluminium) to conduct heat away. Most of them have fins, to allow convection currents in the air to carry the heat away. Also, they are painted matt black so as to radiate heat efficiently. A special heat-conducting paste is applied to the surface in contact with the tag.

However, a transistor usually does not need a heat sink if it is switched fully off (no current) or fully on (saturated). When it is saturated, its resistance to the collector current is very low. A low resistance converts little power to heat.

> **Memo**
>
> $P = I^2R$. Because R is small, then P is small.

Designing a switch

We need a circuit that switches on an LED when the light level falls. It could be the first stage in a security system for detecting intruders. Its system diagram has the typical three stages:

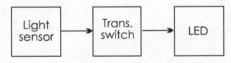

The light sensor can be a voltage divider, made from an LDR and a resistor. The output current from the sensor goes to a transistor switch, consisting of a transistor with a base resistor. This switches an LED with its series resistor. The complete circuit is:

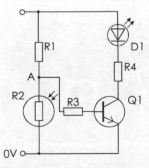

R1 and the LDR (R2) are arranged so that the voltage at point A rises as the LDR receives less light.

The power supply will be a **6V** plug-in PSU because the circuit will be run night and day. The sensor will be the popular and easily available **ORP12**.

A typical LED takes **20 mA** when lit. A low-power transistor such as a **BC548** can switch up to 100 mA, so we will settle on this type for Q1. When Q1 is saturated, there will be almost 6 V across D1 and R4. The forward voltage drop is about 2 V, so we need to drop 4 V across R4. Ohm's Law tells us that R4 must be 4 V divided by 20 mA, which gives **200 Ω**.

Using a multimeter, we may find that the resistance of R2 in low light (the LED-ON level) is 1.3 kΩ. To switch on Q1, we need a voltage at point A that is over 1 V. A few calculations with some of the E24 values show that if R1 is **3.9 kΩ**, then the voltage at A is **1.5 V**. This gives us a margin to allow for resistor tolerance.

When Q1 is on, the voltage at A is 1.5 V (as just calculated) and that at the base is 0.7 V (p. 71). The voltage drop across R3 is 0.8 V. If the gain of Q1 is 100 (p. 71), the base current must be (20 mA)/100, or **200 μA**. So the resistance of R3 must be 0.8 V divided by 200 μA, which gives 4 kΩ. The nearest value is **3.9 kΩ**.

Here is the final design, including the critical voltages and currents:

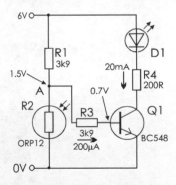

> **Things to do**
>
> Check the calculations by building and testing the circuit.

In the Lab Building on stripboard

Stripboard is a handy way of assembling an electronic circuit. It is particularly useful when you intend to build the circuit only once.

The components are mounted on the plain side of the board. The terminal wires and pins of the components are passed through the holes in the board

The wires and pins are soldered to the copper strips on the rear side of the board.

In this way we build up a circuit with the copper strips as the connecting conductors. We may use wire links to connect one copper strip to another.

Sometimes a copper strip is cut across to separate it into two or more parts.

Stripboard has the important advantage that is is easy to make alterations to the circuit after it has been built. This makes stripboard very useful for building **prototype** circuits. These are the first versions of circuits that have been designed, but not tested in action. After testing, they may need modifying to improve their performance. For example, we may decide to change some of the component values, or to correct the logic. This is almost impossible to do when the circuit is built on a PCB (p. 90).

Tools

- **Junior hacksaw**: for cutting the stripboard to size.
- **Medium file**: for smoothing the cut edges of the board and (optionally) rounding the corners.
- **Magnifier**: preferably 8X or 10X; essential for examining soldering work at all stages.

- **Spot face cutter**: for cutting the strips where a break is needed.
- **Soldering iron**: Low-power (about 15 W), with a fine bit (not more than 2 mm diameter).
- **Wire stripper**: for removing insulation from the ends of connecting wires.
- **Wire cutter**: for cutting connecting wires and for trimming off the projecting ends after component and connecting wires have been soldered into the board.
- **Heat shunt**: to absorb heat and prevent it reaching delicate components such as transistors.

Materials

- **Stripboard:** is sold in standard sizes. It is often more convenient (and cheaper) to buy a large board and cut it into pieces of suitable shape and size.
- **Solder:** Use cored solder, 60% tin:40% lead, preferably 22 swg (0.7 mm diameter).
- **Wire:** For on-board connections: single-stranded (1/0.6) with PVC insulation. It is helpful to have several different colours, such as red, blue, black and yellow. For off-board connections to the control panel, other boards, and power supply: multi-stranded wire, with PVC insulation in different colours. For normal use, 10-stranded wire is best (10/0.12). For special purposes you may need: heavy-duty wire, sheathed wire, computer cable, telephone cable, ribbon-cable or mains cable.

Laying out the board

The first stage is a schematic or circuit diagram of the circuit you intend to build. You may design this yourself or use a design from a book or magazine. The example we are following in this Topic is the 7555 monostable circuit shown on p. 116. You may decide to check it first by building it on a breadboard, as in the photo on p. 117. This gives you a chance to check component values before building a permanent version of the circuit.

Plan the layout of the stripboard on squared paper.

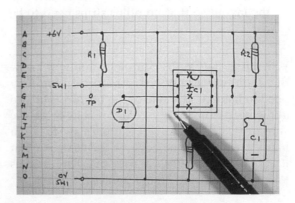

The layout shows:

- **Copper strips**: these are the horizontal lines of the squared paper. The corners of the squares are where the holes are located. To make the layout clearer and to avoid mistakes in the layout, we have inked over the supply line, the input line from S1, and the line from D1 to R3, a connection which might not be clear otherwise.

- **Components**: Allow for the size of the components that you will be using. If in doubt, place the component on a spare scrap of board and count the number of holes required. Normally, a 0.25 W resistor lying flat on the board needs a minimum of four strips *between* the strips its wires are soldered to. Resistors may also be stood on end and can then be soldered to adjacent strips.

- **Terminal pins**: labelled to indicate whether they are power connections or connections to off-board components. In some designs you may use an edge-connector or some other type of connection.

- **Test points**: It makes testing simpler if certain points in the circuit have a terminal pin to which test equipment can be connected. An example is the output of the timer IC at pin 3. The test point is marked 'TP' in the diagram.

- **Wire links**: As far as possible, run these at right angles to the strips.

- **Cut strips**: These are each marked with a bold 'X', preferably with a pen of different colour. In this circuit, we have had to cut the four strips that run beneath the IC. Apart from these, we have been able to arrange the layout so that no other strips need to be cut.

When the diagram is complete, you will know the minimum size of board required. You may need to increase the size of the board slightly to allow space for drilling holes for bolts or plastic stand-offs. Alternatively, you may leave a blank area along one edge of the board so that it can be slotted into a support. This area is labelled 'mounting' in the photo.

Assembling the circuit

Cut the board: If you do not have a ready-cut standard board, use a junior hacksaw to cut a board to the required size. Before you cut it, check that it really will fit inside the box you intend to use. After cutting the board, smooth its edges with the file. Use the file to round the corners slightly.

Cut the strips: Use a spot face cutter for this. When you have finished, use a magnifier to examine the cuts. Look for incomplete cuts, where a 'wire' of copper still connects the strip across the intended cut (A, below). Also look for and remove any flakes of copper that might bridge the gap between adjacent strips (B, below).

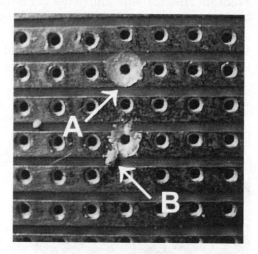

Solder wire links: See p. 79 for soldering procedures. It is usually best to start by soldering one stripped end of the wire into its hole (see below).

Then cut the wire to length, strip its other end, bend it, and insert it in the other hole, pulling it tight. Solder the other end into its hole. Trim both ends.

If there are IC sockets in the layout they can be soldered in place now. They act as 'landmarks' and help you to avoid soldering components in the wrong places.

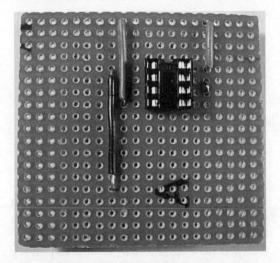

When soldering IC sockets, begin by soldering two diagonally opposite pins. Then check that the socket is lying flat on the board. If not, remelt the solder on one or both pins and push the socket flat. Solder the remaining pins.

Solder resistors: Check values carefully and that their wires go to the correct holes.

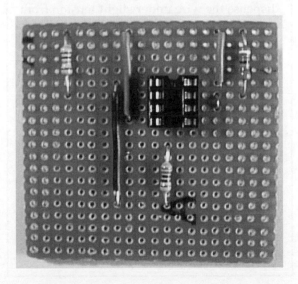

Solder other small components: These include terminal pins, small capacitors, preset pots, diodes, transistors, and sockets (if you have not assembled these earlier). Check that polarised components such as diodes and tantalum capacitors are inserted the right way round.

Solder larger components: These include electrolytic capacitors, which too must be the right way round. Solder leads to off-board components.

Last thing: Remove ICs from their packaging and insert them in their correct sockets (right way round).

Visual check

Using a magnifier, look *again* for badly cut strips, badly made solder joints, and blobs or fine threads of solder causing short-circuits between adjacent strips. Check *again* that the components and links are in the correct holes. Check *again* that diodes, polarised capacitors, transistors and ICs are the right way round.

Refer to page 100 for testing procedures.

Modular approach

The order of assembly outlined above is best for most projects, especially small ones. However, there are some projects where a different order is preferable. If a circuit has several stages it may be better to built it stage by stage. When each stage is built, test it before adding the next stage on to it.

Another way, that is a combination of both techniques is to assemble the whole circuit as described above but to leave out the connecting links between each stage. Then test each stage separately before adding the links to join it to the other stages.

Desoldering

If a component or wire is soldered in the wrong place, it is usually easy to remove it. Simply re-heat the joint with the iron, then pull the wire out. A pair of forceps (tweezers) helps to avoid burnt fingers.

Sometimes this operation may leave too much solder on the strips. This blocks the holes and makes it difficult to insert other components. One way of removing unwanted solder is to use **solder braid**. This consists of fine copper wire braided into a ribbon.

Touch the end of the braid on to the soldered area. Melt a little solder on the tip of a *hot* iron to give good thermal contact. Touch the iron against the braid. The solder beneath the braid melts. The braid acts as a wick and soaks up the molten solder. Remove the braid and, when the solder in it has solidified, clip off the used end of the braid.

Another technique is to use a **solder sucker**. This consists of a nozzle of heat-resistant plastic and a squeezable bulb. Melt the excess solder with the iron, then use the solder sucker to suck it up.

Solder suckers are useful when there is a lot of solder to remove but, with present-day miniature components, there is rarely any need to remove such large amounts.

Tag board — Extension Box 41

This consists of an insulating strip with a row of metal solder-tags on either edge. The strip is supported on one or more brackets bolted to the box or chassis of the project. The tags act as anchor points where components are soldered together, sometimes with wire connections to other tags.

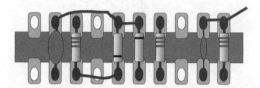

The technique was popular in the days of valve circuits, when components were larger and voltages higher than they are today. It is not possible to mount ICs on tag strip, which greatly restricts its usefulness.

Wire wrapping — Extension Box 42

This technique is used for hand-wiring of circuits that consist mainly of integrated circuits. The wire is insulated with Kynar or Tefzel. Special wire-wrap 'pens' are used to dispense the wire under slight tension from a small reel on the 'pen'. The circuit is assembled on insulated board with perforations spaced as in stripboard, but with no copper strips.

Wire-wrap IC sockets have long pins that are square in section with very sharp corners. When the wire is wrapped around a pin under tension, the corners bite through the insulation and into the wire, making good electrical contact.

The operative runs the wire from one pin to another, making a few turns around each, so making the pin-to-pin connections.

Soldering

Cleaning: Make sure that all surfaces are free from grease. Even handling the board with bare fingers can make it too greasy. If the copper strips or the wires are at all tarnished, rub them with fine emery paper until they gleam.

Heating: Let the soldering iron warm up before you start work. Touch the iron to the end of the solder coil. A *small* amount of solder melts and coats the tip thinly. For best soldering, the combination is 'hot iron plus quick action'. Pass the wire link or component lead through the hole in the copper strip. Touch the tip of the iron against both the wire and the strip at the same time. After a few moments, feed a little solder into the angle between the strip and the wire (Note: on to the wire and strip, NOT on to the iron). With practice, you will learn to feed in enough but not too much. The solder must flow into the angle and wet all surfaces freely.

Finishing: Remove the iron. Do not move the board or wire until the solder has set (about 3 seconds, if you have not over-heated the joint). Use pliers or a wire cutter to cut off the excess wire protruding from the joint.

Inspecting: Use a hand lens to examine the joint. If there is not enough solder, or if the joint is 'dry' (arrowed, right), repeat the soldering. A 'dry' joint is often the result of not having the iron hot enough, or trying to solder on to a greasy surface.

Heat shunt: When soldering heat-sensitive components such as diodes and transistors, clip a heat shunt to the wire leads of the device before soldering (left). The heat from soldering is diverted into the heat shunt instead of travelling up the wires to the device itself. Even with a heat shunt, still try to solder the joint as quickly as you can to avoid over-heating.

WARNING! A hot soldering iron does not look hot, but is hot enough to burn your skin, or your clothes, or the surface of the workbench. It is also hot enough to melt the plastic insulation on wires. Handle the iron with great care!

24 Field effect transistors

These are known as **FETs**, for short. They are active devices (p. 70) and there are several kinds of them, including junction FETs (JFETs) that were widely used in the past. Nowadays an enormous range of n-channel metal oxide silicon FETs (n-channel MOSFETs) is available and, with BJTs, are now the most commonly used types. We deal only with MOSFETs in this book and will simply call them FETs.

There are low-power, medium-power and high-power FETs, enclosed in the same kinds of package as BJTs. The photo shows a typical high-power FET. The case is the same as that of the high-power BJT seen on p. 72, but this photo shows the other side of the transistor.

We can see that the metal tag extends down the body of the transistor. This gives a large area for contact with a heat sink. It also means that the tag is in close contact with the actual transistor in the body of the device.

The photo also shows that the FET has three terminals. These are named **source, drain** and **gate**. They roughly correspond with the emitter, collector and base of a BJT, but there are important differences. The most important practical difference is that virtually no current flows into the gate of an FET.

In normal use, an FET is connected in a similar way to a BJT. The source is the most negative terminal and the drain is the most positive. When a positive voltage is applied to the gate, a current, the **drain current** flows in at the drain and out at the source. The next discussion looks more closely at this.

Transistor action

An FET can be used as a transistor switch, as in this circuit:

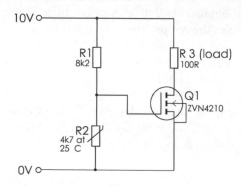

The transistor switches on the load when the temperature falls below a given level. The transistor is switched on by the *voltage* output of the voltage divider. As temperature falls, the resistance of R2 increases. When the voltage across R2 exceeds the **threshold voltage** of the FET, the transistor begins to turn on. The threshold voltage varies with different types of FET, but is often between 2 V and 4 V.

Once the threshold voltage is exceeded, further increase of voltage may rapidly saturate the transistor. Here is a graph of the drain current (I_D) through a 100 Ω load against the gate-source voltage (V_{GS}):

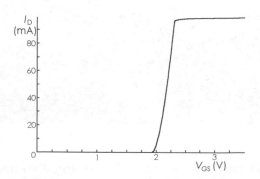

As V_{CS} increases above the threshold (2 V), the current increases rapidly. The FET saturates at about 2.3 V.

The FET saturates because the voltage across the load resistor brings the drain terminal almost down to zero. With a smaller load, the voltage drop is smaller and we can see the shape of the voltage-current curve more easily:

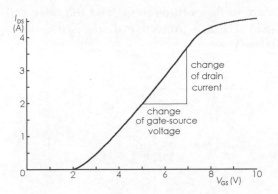

Compare this graph with the graph for base current and collector current of a BJT (p. 70). The line is reasonably straight in the middle, so that current is proportional to voltage. However, it is not straight at the ends, which means that current is only approximately proportional to voltage if the signal voltage varies over a wide range.

What might be called the 'gain' of the FET is the change of drain *current* for a given change in gate-source *voltage*. In the central region of the graph above, the current increases by about 1.7 A when the voltage increases by 2 V. We can say that the 'gain' is:

$$1.7/2 = 0.85 \text{ amps/volts}$$

We have already said (p. 28) that volts/amps is called resistance. The inverse, amps/volts, is known as **conductance**. In the case of an FET, where the voltage is the *input* and the current is the *output*, we call this the **transconductance**, g_m, of the FET:

$$g_m = \frac{\text{change of drain current}}{\text{change of gate-source voltage}}$$

The unit of conductance and transconductance is the **siemens**, symbol **S**. The FET illustrated by the graph above has a transconductance of 0.85 S, or 850 millisiemens.

BJTs and FETs

BJTs and FETs have similar actions, but distinctive properties:

- **Conversions:** BJTs convert current to current. FETs convert voltage to current.
- **Input current:** A BJT needs an input current. An FET requires no input current.
- **Input/output:** The input/output relationship of a BJT is linear (a straight-line graph), but that of an FET is not linear for large signals. This may lead to distortion of large signals by an FET.
- **Speed:** FETs switch faster than BJTs, though both types are fast enough for most applications.
- **Input voltage:** An FET switches on when the gate-source voltage exceeds the threshold voltage. The gate voltage can be any voltage between the threshold and the supply voltage. when the FET is on. The base-emitter voltage of a BJT remains close to 0.7 V when it is on, whatever the size of the input current.
- **Input resistor:** An FET does not require a resistor in its gate circuit. This simplifies circuit-building.
- **Output resistance:** Most FETs have very low resistance when switched on, usually less than 1 Ω. This makes them good transistor switches.

Switching a relay

Sometimes we want to use a transistor switch to control a high-power device. It may require larger current or higher voltage than a power transistor can handle. Or we may want to switch an alternating current, which is something that transistor switches can not do. This is when we use a relay (p. 55)

In this example, the system has four stages. The sensor is a thermistor. This controls an FET switch, which switches a relay coil. The relay contacts control a mains-powered heater.

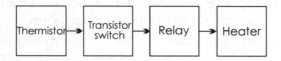

As in many switching circuits, the first stage is a voltage divider. There is a variable resistor to set the switch-on temperature. There is no resistor between the divider and the gate of the FET.

When the temperature falls, the resistance of R2 increases. The voltage at the wiper of VR1 increases. Q1 is switched on when this voltage exceeds the threshold voltage. Current flows through Q1 and the relay coil (RL). D1 is a protective diode (see Box 44, below).

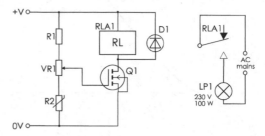

The relay switches a separate mains circuit, using its normally-open contacts. When current flows through the relay coil, the normally-open contacts close. The heater circiit is completed and the heater comes on. It stays on until the temperature reaches the set level.

Protective diode

A protective diode is essential when switching a load that has high inductance. Inductive loads include relay coils, electric bells and electromagnets. The diagram on the right illustrates the effect of self-induction (p. 46) in the coil.

(a) When the transistor (BJT or FET) is on, current is passing through the coil. Possibly the current varies slightly and, if so, the magnetic field varies slightly. But there are no dramatic changes.

(b) When the transistor switches off, the current ceases *instantly* and the field collapses *rapidly*. The effect of induction depends on the *rate of change* of the field. The faster the change, the greater the effect.

Total switch-off is a very fast change. A very large electric force is induced, to try to keep the magnetic field as it was. This force may be equal to several hundred volts, even though the original voltage across the load was only, say, 10 V. A current of several amps surges through the transistor and burns it out.

(c) The solution is to wire a diode as shown, to conduct the excess current safely away.

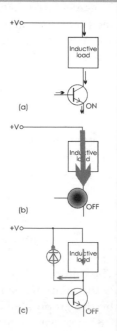

Data sheets
Extension Box 45

There are so many thousands of different types of transistor and other components that no electronic engineer is expected to remember their characteristics. Yet such information is essential when you are planning a new project. Here are some of the sources that you can use.

Manufacturers' data sheets

These are the most detailed source of information. They describe what a device does and list possible applications. There are tables that tell you, for example, the absolute maximum power supply voltage that you can use, and the highest frequency at which it can operate. Other tables specify how the device operates: what size of voltages and currents it responds to and what output voltages and currents it delivers. The sheets often include notes on how to use the device, and suggest circuits that you might find useful when designing a project.

But manufacturers' data sheets can be hard to use. They contain so much information that it may be difficult to pick out the item you need. Also, they are written for the electronics professional. The jargon may be difficult to understand.

Finally, the data sheets describe only the products of a particular manufacturer. You are not able to compare products of different manufacturers, except by building up a collection of sheets from several manufacturers.

Internet

Most component manufacturers have a web site to promote their products. They publish their data sheets there, often in downloadable form. They may also produce CD-ROMs that carry the same information as the web site. This is an advantage if you prefer to browse the data sheets off-line.

These CDs are often available free or at a very low cost.

Continued overleaf

Data sheets (continued)

Suppliers' catalogues

These are probably the most useful sources of information for the student. The data tables in the catalogues are usually limited to telling you the things you are most likely to want to know. They list the products of all the more important manufacturers, which makes it easier to pick out the type that best suits your needs. The bigger suppliers have web sites and may issue their catalogue as a low-cost CD-ROM.

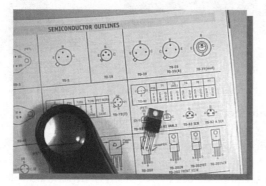

Electronics magazines

These publish details of the newest components, and articles which explain how to use them. They are a valuable source of ideas for projects. Occasionally, the magazines publish free data sheets or CD-ROMs, that provide much useful information.

Books

Data books can be useful because they deal mainly with the most popular devices. They may include advice that helps you choose which type to use. The main drawback with such books is that they eventually become out of date. The newest devices are not mentioned and some of the devices listed in the book may no longer be available.

Selecting transistors

For most components, the data sheets and catalogues are fairly easy to use. For example, when choosing a resistor the resistance and tolerance are usually the only things that matter. In some applications we may also need to consider power rating too. For transistors, there are several features that need to be thought about when selecting a suitable type.

For your examination work, you are concerned with only two classes of transistor. One is the npn bipolar junction transistor (or BJT). The other is the n-channel MOSFET (or FET). The comparisons of the two classes on p. 81 will help you decide which is better for your circuit. If you decide on a BJT, study Extension Box 47, opposite.

Selecting a BJT (1)

This table is set out like a typical data sheet for BJT transistors:

Type	Case	Ic (max)	Ptot	hfe at Ic (mA)		Ft at Ic (MHz at mA)		Applications
BC107	TO-18	100 mA	300 mW	110-450	2	300	10	GP small signal amplifier
BC108	TO-18	100 mA	300 mW	110-800	2	300	10	GP small signal amplifier
BC109	TO-18	100 mA.	300 mW	200-800	2	300	10	Low noise, small sig. amp.
BC109C	TO-18	100 mA	300 mW	420-800	2	300	10	Low noise, high gain, amp
BC337	TO-92	500 mA	625 mW	100-600	2	100	10	Output stages
BC548	TO-92A	100 mA	500 mW	110-800	2	300	10	Low noise, small sig. amp.
BC639	TO-92	1 A	1 W	40-250	150	200	10	Audio output
BD139	TO-126	1.5 A	8 W	40-250	150	250	50	Power GP
BD263	TO-126	4 A	36 W	750	1.5 A	7	1.5 A	Med. power Darlington
BUX80	TO-3	10 A	100 W	30	1.2 A	8		High current switch
MJE3055	TO-220	10 A	75 W	20-70	4 A	2	500	Power output
2N2222	TO-18	800 mA	500 mW	100-300	150	300	20	High speed switch
2N3641	TO-105	500 mA	350 mW	40-120		250	50	GP, amplifier, switch
2N3771	TO-3	30 A	150 W	15-60	1.5 A	0.2	1 A	Power output
2N3866	TO-39	400 mA	1 W	10-200	50	500	50	VHF amplifier
2N3904	TO-92	200 mA	310 mW	100-300	10	300	10	Low power GP

The transistors are listed by **type number**. Many of the circuit diagrams in this book specify type number BC548. This is an easily available, cheap, general-purpose BJT, made by several different manufacturers. As can be seen from the table, there are several other types of BJT that would work in these circuits just as well as the BC548.

The type number is not particularly useful when you are selecting a transistor. The easiest way is to refer to the final column of the table and select by application. The majority of transistors are described as 'general-purpose' (GP) possibly with some additional features mentioned. One of these types will suit most projects.

If you have a special application in mind, look for descriptions such as 'switching transistor', 'audio driver' , 'power switch', 'radio-frequency', and 'high gain'.

The table lists the **case** of the transistor. TO-18, for example, is the small metal case seen in the photo on p. 68.

The TO-92 plastic case is in the same photo. The TO-3 is the power transistor case (right).

Selecting a BJT (2)

This box explains how to use transistor data tables when you are selecting a transistor to suit a special application.

Probably the most important item is the **maximum collector current**. The data sheet lists this in a column headed I_C **(max)**. This is the maximum collector current (almost equal to the emitter current) that the BJT can pass when it is saturated. It is equal to the current passing through the load.

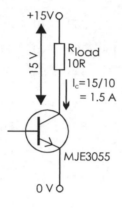

As an example, we will select a transistor for a transistor switch circuit. The supply voltage is 15 V, and the resistance of the load is 10 Ω. When the transistor is saturated, there is almost 15 V across the load. The current through the load is 15/10 = 1.5 A.

The only transistors in the data table that are able to carry 1.5 A are the BD139, BUX80, MJE3055, and 2N3771. The BD139 can only *just* carry the required current. Select one of the others, say the MJE3055, which is the cheapest of the three.

In a switch circuit, the transistor is sometimes off, with no current flowing through it and no power being converted to heat. At other times it is fully on, with only a small voltage across it. The power is low and little heat is produced.

The situation is different in some kinds of amplifier, because the transistors are never saturated. On average, the transistor is half on. With a 15 V supply, the voltage across it is 7.5 V. Given a 4 Ω speaker as the load, the current is 7.5/4 = 1.875 A. Electric current is being converted to heat in the transistor at the rate of $7.5 \times 1.875 = 14$ W.

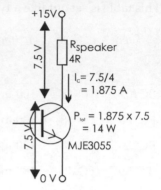

In the table, the **maximum power** (P_{tot}) rating of the BD139 is 8 W, so it can not be used here. Of the other power BJTs, those in TO-3 cases are usually dearer.

Choose the MJE3055, which is in a cheaper cases.

The **current gain** specified in the table is the small signal current gain, h_{fe}, defined on p. 71. It is measured with the collector current at the stated value. Usually, a range is specified because transistors of the same type may vary widely in gain. The BC548, for instance, has a gain of 110 to 800. Sometimes transistors are graded by the manufacturer and sold in gain groups. A transistor sold as a BC548A, for instance, has a gain in the range 110 to 220. BC545B ranges fron 200 to 450, and BC548C ranges from 420 to 800.

The fourth important feature of a transistor is its **maximum frequency**, F_t. A transistor being used as a switch responds quickly enough for most applications. When working at radio frequencies, it may not be able to respond quickly enough to the alternating currents. The gain falls off at high frequencies. F_t is defined as the frequency for wich the gain is 1. For many transistors, F_t is rated in hundreds of megahertz and, in the table, is quoted in megahertz at a given collector current.

Normally, we do not have to worry about F_t. It is mainly in radio circuits and computer circuits that such high frequencies are used. But it can be seen from the table that high-power transistors have much lower F_t than other types.

Questions on sensors and transistors

1 What is a light dependent resistor? What is its most important property?

2 Describe how to build a light meter, using a light dependent resistor, a multimeter, and any other components you need.

3 Draw a system diagram of the light meter that you described in Q. 2.

4 Describe a thermistor that you have worked with. What is its most important property?

5 What is meant by the term 'negative temperature coefficient'?

6 Name the three terminals of a BJT. Out of which of these does the current flow when the BJT is connected into a circuit?

7 Describe how you would demonstrate transistor action and what results you would obtain.

8 Which is the biggest, collector current, emitter current, or base current?

9 Draw the symbol for a push-to-make push-button.

10 A transistor switch circuit uses a BJT to switch on an LED when the temperature is greater than a set value. Draw a system diagram of the circuit.

11 Draw a circuit diagram of the circuit of Q.10 and explain how it works.

12 A transistor switch is being used for switching an LED. The supply voltage is 12 V. Calculate a suitable value for the series resistor.

13 What do we mean by the term 'small signal current gain'?

14 Describe how you would measure the small signal current gain of a BJT.

15 When the base current of a BJT is 40 µA, its collector current is 35 mA. When the base current is 65 µA, the collector current is 38 mA. Calculate its small signal current gain.

16 What is the typical small signal current gain of a BJT? What is its base-emitter voltage when it is switched on?

17 What is meant when we say that a BJT is saturated?

18 Why does the transistor of a transistor switch not usually need a heat sink?

19 Design a transistor switch to be used with an ORP12 as sensor, to switch a 12 V 6 W lamp when the light level increases above a set level.

20 What are advantages of FETs over BJTs when used as switches?

21 What do we mean by the threshold voltage of an FET?

22 What is the difference between gain and transconductance?

23 An FET has a transconductance of 3 S. Its gate voltage is above the threshold and increases by 1.5 mV. What change occurs in the drain current?

24 Draw the symbol for an FET and name the terminals.

25 Name two passive and two active devices.

Extension questions

26 Design an FET switch circuit to turn on a 230 V, 100 W lamp in a porch at night.

27 Design a touch-switch circuit (p. 72) to activate a 12 V, 1.5 A door catch release in a security system.

28 Using the data table on p. 85, select a transistor to use in the following circuits:

(a) a switch circuit for an LED running at 20 mA when the base current can be no more than 40 µA.

(b) a switch circuit for a DC motor running on a 3 V supply at up to 1 A.

(c) the output stage of an audio amplifier running on an 18 V supply, with an 8 Ω speaker.

(d) a switching circuit in a computer running at 150 MHz.

29 List five sources of transistor data to which you have access. Which is the most useful and why?

25 The structure of a system

An electronic kitchen scale is a simple example of a system.

It consists of:
- a force sensor,
- a microcontroller for taking a signal from the sensor and producing signals to send to the display,
- a display.

Here is the system diagram:

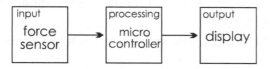

The system has three parts, just like the systems drawn on pages 69 and 73. In the diagram above we have named the parts as force sensor, microcontroller and display. We can understand the *system* without knowing anything about force sensors, microcontrollers or displays. We know what each part does, even though we do not know how it does it. This is described later in the book.

The diagram above also names the parts of the system in more general terms: input, processing and output. We can apply these names to many different systems. On p. 69, for example, the LDR provides the input, the transistor switch does the processing , and the lamp indicates the output.

Remember that the arrows in the system diagram do not represent wires carrying currents. They represent a **flow of information** through the system. At each stage this information is represented by forces, voltages or currents until, in the output stage, it is represented by numerals on the display.

The sliding doors of a shopping centre are opened and closed by a system that has a similar structure.

The system diagram is:

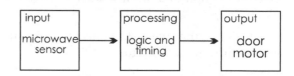

As before, there is a flow of information through the system. This time the information is 'A person is approaching the door'. The sensor detects the pattern of reflected microwaves that represents this information. The information is then converted into electrical signals. Finally the information is acted upon and the door motors are turned on.

A more complicated system is needed to run a supermarket checkout.

It has four inputs: the bar code scanner (1), a keyboard for use by the checkout assistant (2), an electronic scale for weighing vegetables (3). and an input from the in-store computer which holds the prices of the goods.

The checkout has four outputs: a display for the checkout assistant (4), a display for the customer, a printer to print the invoice, and an output to the in-store computer to update the stock list.

A small computer is at the centre of the system to process the information from the inputs and send information on to the outputs.

Things to do

Draw system diagrams of (a) a supermarket checkout, (b) an automatic teller machine, and (c) any other electronic system that you know about.

This is an Automatic Teller Machine (ATM). It has two inputs: a card reader (1) and a small keypad (2). It has three outputs: a screen (3), a receipt printer (4), and a banknote counter (5). It has a modem to connect it by telephone line to the bank. Processing is done by its own computer.

The ATM can give you cash drawn from a bank account almost anywhere in the world.

Designing and building systems

This part of the book describes components and circuits that are useful in building systems. The topics are listed below, under the headings Input, Processing and Output.

Use this part as a source of ideas and practical help for the systems and projects you build. Learn about other systems that you do not have time to build.

Input

26 Sensors for temperature, light, force, sound, magnetic field, position, vibration, moisture. Switching sensors.

27 Interfacing sensors Ways of connecting sensors to systems.

Processing

28 Amplifying signals

29 Timing

30 Logic The basic logic gates

31 Logical sequences

32 Storing data

33 Microcontrollers

34 Programs

Output

35 Visual output Lamps, LEDs, displays. Data output. Transducers.

36 Audible output Bells, buzzers, sirens, loudspeakers, sounders.

37 Mechanical output Motors, solenoids.

Printed circuit boards

PCBs were invented for mass production of electronic equipment. The idea is that a plain circuit board is 'printed' with a copper film that provides all the connections between components. If the master layout is correct, all PCBs made from this will be correct. All that is necessary is to drill holes for the component wires or pins, insert the components, and apply a wash of molten solder. The whole process, including most of the placing of components, can be automated. Large numbers of circuit boards can be produced at greater speed and lower cost.

PCB designs are popular with many constructors because they make circuit-building easier and quicker. There is less chance of making mistakes. Designing and making your own PCBs helps you to understand the techniques used in the electronics industry. It gives you an idea of the economics of industrial production techniques.

PCB techniques

A PCB begins as a board of insulating material coated on one or both sides with a layer of copper. We will not refer again to double-sided boards, but the techniques for handling them have much in common with those used for single-sided boards.

The principle of making PCBs is to prepare a **mask** that shows the copper **pads** to which the component wires are soldered. The mask also shows the **tracks** that join the pads together. The tracks are the equivalent of the copper strips and wire links that we use in the stripboard technique.

In various ways, to be described later, the mask is applied to the copper surface of the board. Then the board is **etched**. This dissolves away the copper, except where it is protected by the mask.

This leaves the copper pads connected by tracks. Before the components can be mounted on the board, a hole is drilled in each pad to accept the component leads. This stage is omitted when assembling a board with surface-mount components.

From then on, the technique is very much the same as with stripboard. Components are soldered in place, beginning with wire links and resistors and finishing with the largest components. The board is visually inspected and finally tested.

Tools

Most of the tools required for PCB production are the same as those used for stripboard and described on p. 75. The only one that is *not* needed is the spot face cutter.

Additional tools and equipment required are:
- **UV exposure box**: for use in photo-resist methods (see opposite)
- **Shallow glass or plastic dish**: for etching, or an **etching tank** (preferably with an air pump and heater).
- **Plastic forceps**: for handling PCBs during etching.
- **Goggles and gloves**: for protection from corrosive fluids used for etching.
- **Abrasive block**: or equivalent for cleaning etched board.
- **Mini electric drill**: with 0.8 mm and 1 mm bits, and possibly a few of larger diameter.

Materials

- **Blank PCB board**: with a layer of copper on one side. If you are using a photo-resist technique it is best to use board already coated with the photosensitive resist layer. It is also possible to obtain spray-on resist.

- **Resist pen**: For masking on hand-drawn boards.
- **Rub-down transfers**: For masking hand-made boards. Obtainable as sheets of transfers of various kinds such as round pads of various diameters and IC pads.
- **Acetate sheet (thick)**: Required for the photo-resist techniques. The thicker kind is used as a base for a mask made from rub-down transfers.

- **Acetate sheet (thin)**: The thinner kind is used with a laser printer. This has a paper backing that is peeled off after printing on to the acetate sheet.
- **Etchant**: This is usually a solution of ferric chloride. It is highly corrosive and must be handled with great care. Always wear goggles and gloves when handling it. It can be re-used several times.

Laying out the circuit and making the mask

The diagram shows the various techniques. More details are given on the next two pages.

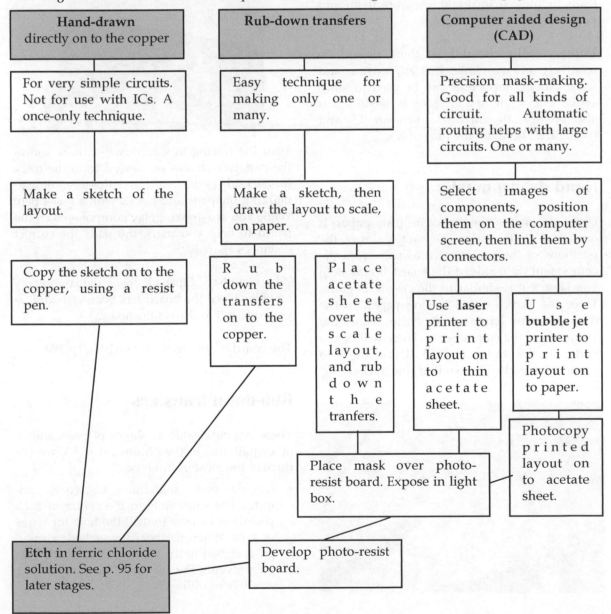

Hand-drawn directly on to the copper	**Rub-down transfers**	**Computer aided design (CAD)**
For very simple circuits. Not for use with ICs. A once-only technique.	Easy technique for making only one or many.	Precision mask-making. Good for all kinds of circuit. Automatic routing helps with large circuits. One or many.
Make a sketch of the layout.	Make a sketch, then draw the layout to scale, on paper.	Select images of components, position them on the computer screen, then link them by connectors.

Copy the sketch on to the copper, using a resist pen.

R u b down the transfers on to the copper.

P l a c e a c e t a t e s h e e t over the s c a l e layout, and rub d o w n t h e tranfers.

Use **laser** printer to p r i n t layout on to thin a c e t a t e sheet.

U s e **bubble jet** printer to p r i n t layout on to paper.

Photocopy p r i n t e d layout on to acetate sheet.

Place mask over photo-resist board. Expose in light box.

Etch in ferric chloride solution. See p. 95 for later stages.

Develop photo-resist board.

91

.Layout

We now look at the stages of PCB production in more detail. As an example, we use the same 7555 monostable circuit that we used to illustrate the stripboard technique on pp 74-9.

The first step is to plan the layout of the components on squared paper. Components should be aligned parallel with the edges of the board. Arrange the components so that they are not too widely spaced, but not so close together that they are difficult to get at. You need to know the distances between the wire leads or pins. If in doubt, measure them and keep a note of the distances.

With a well-planned PCB there should be relatively few wire links. Instead, resistors and some other components can be used to jump across tracks where a crossing is needed. It is also possible to run tracks beneath ICs and (with care!) between the pins of ICs.

Hand drawn masks

Draw a sketch of the circuit on plain paper. It need not be to scale. The sketch shows the positions of the components as seen from the component (top) side of the board. It may help you later if it is similar to the circuit diagram. Trace the sketch on to tracing paper. In the tracing, draw circular spots to show the terminals of the components. These will be the pads on the finished PCB. Draw lines to connect the pads. These will be the tracks.

Cut the board to size, using the junior hacksaw. Smooth the edges with a file.

Clean the board by scrubbing it with a scouring pad or use a domestic scouring fluid. Rinse the board and dry it with a paper tissue. Try not to touch the copper surface with your fingers from now on, until it is etched.

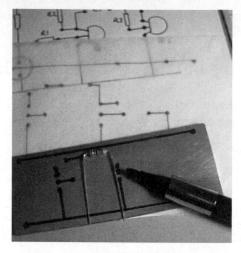

Turn the tracing upside down. It now shows the pattern of tracks as viewed from the track (under) side of the board. Use this as a guide to marking the pads and tracks with a resist pen. While you are drawing, lay components on the board to help you mark the pads the correct distances apart.

When you have finished, check the layout *very carefully*. Once the board has been etched, it is very difficult to correct mistakes.

The board is now ready for etching (p. 95).

Rub-down transfers

These are obtainable as sheets of pads and of other patterns. In the photos on p. 93, we see three of the most useful types:

- Circular pads, sometimes known as 'do-nuts'. The small hole in the centre of each pad shows where to drill the hole for wires or pins. When the board is etched, a small pit is etched in the centre of each pad. This pit prevents the bit from jumping to one side when drilling.

DIL patterns: These have the standard spacing of the pins of double-in-line integrated circuits. **Tapes**: Reels of tape of various widths are used for laying down tracks.

All the rub-down transfers are self-adhesive. When firmly pressed on to the surface of the copper, they act as resist. They prevent the copper under them from becoming etched.

You may need to use a resist pen for drawing parts of the mask that are not easy to produce with transfers.

The first stage in making a transfer mask is to draw a rough sketch of the proposed layout. Check it, and try to reduce the number of wire links required by re-routing some of the tracks. Finally, make a full-size scale drawing, as seen from the component side of the board. Preferably, make this on thin paper or tracing paper.

Cut the board to size, using a junior hacksaw. Smooth the edges with a file. Clean the board by scrubbing it with a scouring pad or use a domestic scouring fluid. Rinse the board and dry it with a paper tissue. Try not to touch the copper surface with your fingers from now on, until it is etched.

The next step depends on whether you are going to:

- use the transfers as resist elements and stick them directly to the board, or
- make a mask on acetate sheet and use board with a photo-resist.

The first alternative provides for only one board. With an acetate mask, you can make as many as you want.

Using transfers directly: If the scale drawing is on thin paper, simply turn it over, face-down on the board. If not, copy the scale drawing on to tracing paper. Place this face down on the board.

Turn the tracing over, face-down on the board. Use a sharp pin or needle to prick through the centre of each pad. You need mark only the four corner pins of ICs, as the DIL transfers have the pads correctly spaced. The same applies to transfers of edge-connectors.

Remove the tracing paper and rub down a pad where each pin-point shows on the surface of the copper. The pads are 'landmarks' for laying out the track. Use a sharp knife for cutting the roll of track transfer to the lengths required. Press it firmly in place.

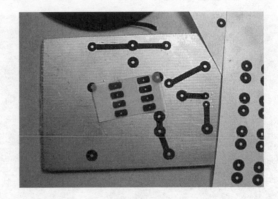

Check everything carefully, as it is very difficult to correct mistakes after etching.The board is now ready for etching (p. 95).

The alternative is to make the mask on acetate sheet. Cut a piece of acetate sheet a little larger than the board size. Place this over the scale drawing. Use a bulldog clip to prevent the sheet from slipping. Rub down the transfers on to the sheet.

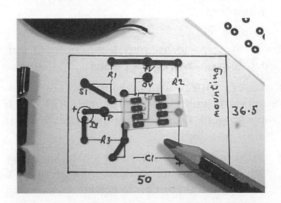

Note that the mask on acetate sheet is the *same way round* as the scale drawing. It is not reversed left-to-right, like the mask made directly on the copper.

Check everything carefully, as it is very difficult to correct mistakes after etching. The board is now ready for etching (see opposite).

CAD

Using CAD software, build up the layout on the computer screen.

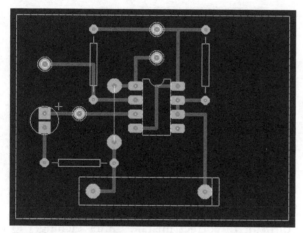

Points to notice about this example include:

- A track runs *under* the IC to link pins 4 and 8 (compare with stripboard layout, p. 74).

- Polarities of the LED and capacitor are indicated.

- Only one wire link used, from pin 1 to R3 and C1.

- Clear area of board on the right, for mounting.

There are two techniques for printing it out, depending on the type of printer:

- **Laser printer**: print on to thin acetate sheet, specially made for laser printers. This has a backing of thin white paper. Peel this off after printing.

- **Bubble-jet printer**: print on to ordinary white paper. Then make a full-size copy of the printed layout on to acetate sheet, using a photocopier.

The printout shows only the pads and tracks:

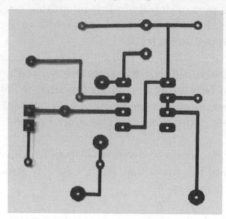

Exposing and developing

This is for masks made by rubbing down transfers on to acetate sheet, or printing on to acetate sheet. You need copper-clad board that is coated with a layer of photo-resist. Before use, the photo-resist is protected from light by a stick-on film of light-proof plastic.

The following operations should be done well away from sunlight or bright windows. Without removing the light-proof film, use a junior hacksaw to cut a piece of board to size.

Peel the light-proof film from the board. Place the mask in contact with the photo-resist. Make sure that it is the right way round. The printed side or the side with transfers must be in contact with the resist.

Place the mask and board in the exposure box. Close the lid and switch on. The exposure time is the standard one used in your lab.

As soon as the board is exposed, remove the mask and develop the board. Use a shallow plastic dish containing photo-resist developer. The developer is a strong caustic solution, so handle it with care. Wear gloves and goggles, and use plastic forceps. Rock the dish gently to keep the solution stirred. When the layout shows clearly on the copper, rinse the board with water.

Etching

The etchant is usually a solution of ferric chloride. This is strongly corrosive, so handle it with great care, using the same precautions as for the developer. An alternative etchant is ammonium persulphate solution.

Boards prepared by any of the processes describe on pages 92 to 94 may be etched in a shallow plastic dish, as used for developing photoresist. Alternatively etch the boards in a special etch tank. A tall narrow tank allows a large board to be etched in a relatively small volume of solution. An air pump blows bubble through the solution to stir it and help the etchant to act evenly. There may be a heater to warm the solution and reduce the etching time.

The board or boards to be etched are held in a wire-mesh basket or suspended from clips. Lift the board every few minutes to see how etching is progressing. Nothing may appear to be happening at first while copper in the exposed areas of the board is becoming thinner. Then the copper in these areas clears away. The copper may take longer to clear between tracks that are close together. Continue the etching process until the tracks and pads are clearly separated, with no incompletely etched bridges between them. Then remove the board from the etchant and rinse it thoroughly with water.

Cleaning

Use an abrasive pad to remove the mask from the board. Smooth the edges of the board with a file, if you have not already done this.

Drilling

A small low-voltage electric drill is used for this. It can be hand-held.

Use a 1 mm bit for terminal pins, and for the thicker component wires. For IC pins and thinner component wires, use a bit of narrower diameter, such as 0.8 mm.

Assembly

The final steps of placing the components, soldering, inspection and testing are as described on pp. 76-8.

Here is the completed circuit:

26 Sensors

Temperature

The sensor for temperature is the **thermistor**, described on p. 66. Its resistance decreases with increasing temperature. The best way to use a thermistor is to connect it in a voltage divider. Then the information about temperature appears as a voltage at the junction of the divider. Putting it another way, the temperature is represented by a **voltage signal** from the divider.

Light

The resistance of a **light dependent resistor** (p. 64) decreases with increasing light. As with the thermistor, it is best used as part of a voltage divider, that produces a voltage signal.

A **photodiode** has the properties of an ordinary diode and is sensitive to light.

In the photo on the right, the photodiode is housed in a metal can. The diode is visible through the lens at the top of the can, as a square chip of silicon.

The photodiode on the left is housed in an opaque plastic case. The case is transparent to infrared light. This diode is useful in security systems to detect the breaking of an infra-red beam that is invisible to an intruder.

A photodiode is connected so that it is reverse biassed. Only a small leakage current of a few microamps passes through it. The current is porportional to the amount of light falling on the photodiode. The current passes through a resistor and a voltage develops across it.

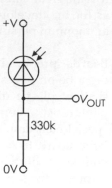

The voltage (V_{OUT}) is proportional to the amount of light.

Force

A **strain gauge** is sensitive to mechanical force. It consists of thin metal foil etched to form a number of very fine wires. This is embedded in a plastic film.

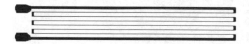

The gauge is cemented to the object that is to be stressed. When the object is stressed there is tension on the foil which stretches the wires. They become longer and thinner so their resistance increases. The change in resistance is very small, so a special circuit is used to measure it.

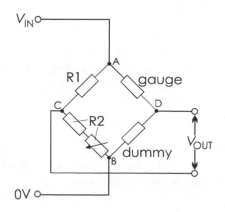

The circuit is a **Wheatstone Bridge**. One arm of the bridge is the gauge and another is an identical gauge, the dummy. This is not put under strain but is included in the bridge to compensate for changes in resistance of the gauge and its leads due to temperature. R2 is a fixed resistor and variable resistor in series.

Think of the bridge as two voltage dividers (ACB, ADB) side by side. The voltage at C is proportional to R1/R2. The voltage at D is proportional to the resistance of the gauge divided by the resistance of the dummy.

One way to read out the result is to adjust the variable resistor until the voltage at C equals the voltage at D. When this is done, the bridge is said to be balanced and V_{OUT} is zero. Then we calculate the resistance of the gauge using this equation:

$$\frac{R1}{R2} = \frac{R_{gauge}}{R_{dummy}}$$

R1 and R2 are known. The resistance of the dummy at a standard temperature is taken from a data sheet, so we can calculate the unknown resistance of the gauge under strain. The final step is to calculate the force from the change in resistance of the gauge. Usually the circuit is calibrated by applying known forces, measuring the changes in resistance, and plotting a graph to relate force to resistance.

A **load cell** consists of one or more strain gauges cemented to a metal bar or ring. The load cell is calibrated by the manufacturer. They are designed to measure tension, compression, or twisting forces. When the bar or ring is strained, the voltage at its terminals is used to find the force. Special electronic devices automatically calculate and display the force on the load cell. Load cells are often used for weighing. Heavy-duty types can be used for weighing hundreds or thousands of kilograms In a weighbridge, they are used to weigh heavily loaded vehicles. Smaller versions are made which weigh masses of a few kilograms.

Sound

Sound is detected with a **microphone** such as a **capacitor microphone**.

For project work, it is often cheaper to use a 'microphone insert', as illustrated in the photo. This is the basic microphone without case and stand. It has two solder pads behind.

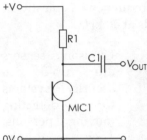

The sound quality from a capacitor microphone is very good, but the voltages it produces are small. Often there is a small amplifier built in.

The complete unit is connected as shown in the diagram above. The resistor required depends on the supply voltage. Check the values on the data sheet for the microphone. Usually the signal from the microphone is passed through a capacitor to the next stage of amplification. This is because a direct connection might draw so much current that the amplitude of the voltage signal would be reduced.

A **crystal microphone** produces a voltage output signal without the need for a power supply. The photo shows a crystal 'microphone insert'.

Often the microphone has a metal case intended to shield it from stray magnetic fields. This case is connected to one of the two terminals. To make use of the shielding effect, always connect that terminal to the 0 V supply line.

The other terminal is connected to the amplifying stage by a capacitor. The reason for this is the same as for the capacitor microphone.

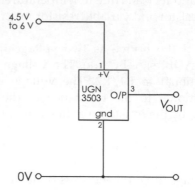

Crystal microphones are cheap but the quality of their output is poor.

An **ultrasonic sensor** is a type of crystal microphone. It is specially made to be sensitive to sound of high frequency. Typically it responds best to sounds at 40 kHz.

Ultrasonic sensors are used in security systems for detecting moving persons. They can also be used to detect when stationary objects are near.

Another application is in equipment for measuring distances, such as electronic tape measures and range finders.

Magnetic field

A **Hall effect sensor** responds to the intensity of the magnetic field around it. It has three terminals. If there is no magnet close to it, its output voltage is about half the supply voltage.

If the south pole of a magnet is brought close to it, the output voltage rises. The amount of rise is in proportion to the strength of the field. If the north pole of a magnet is brought close to it, the output falls.

The UGN3503 is typical of Hall effect sensors. It requires a supply voltage between 4.5 V and 6 V.

It can be used as a switch, simply by moving a magnet towards it or away from it. It has the advantage that it changes state more quickly than a mechanical switch, and needs less force to operate it. It can operate at high speed, switching on and off thousands of times per second.

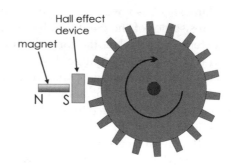

The device can measure the speed of rotation of an axle. There is a toothed iron wheel on the axle. A permanent magnet produces a field through the Hall effect device. This field changes as each tooth passes by the device. The output alternates at a frequency depending on the rate of rotation and the number of teeth on the wheel. A circuit measures the frequency and from this we can calculate the rate of rotation.

Moisture

Moisture sensors are generally home-made. A **water level sensor** has two probes of thick copper wire mounted close together.

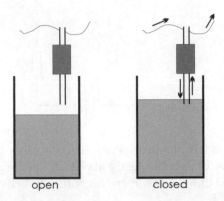

Normally the resistance between the wires is so high that it is like an open switch. When water partly covers the probes, conduction occurs and the resistance between the probes is low. Current flows, as through a closed switch.

Another type of moisture sensor can be made from a small rectangle of stripboard.

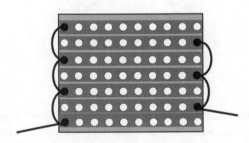

Connect alternate strips by soldered wires. This type of sensor is useful for detecting rain or water sprays.

Position sensors

Position sensors are available but are very expensive. A microswitch (p. 52) can be used to detect whether an object is in position or not. For example, it can tell us whether a cupboard door is open or closed. Better, use *two* microswitches, one for 'door open', and another for 'door closed '.

If you want to determine the exact position of an object, it may be possible to connect it to the wiper of a slider pot (p. 32). As the object moves, the wiper moves. The resistance changes between the wiper and one end of the pot. In this way position is converted to current. The information can then be processed by a logic or amplifying circuit. A rotary pot can be used in a similar way to sense rotary position or angle.

Another way to sense position is to use a beam of light. Preferably use infrared. Direct the beam so that it is broken when the object reaches a certain position. The voltage output from an infrared photodiode sensor (p. 96) falls when the object is in position. Below, a beam is being used at a supermarket checkout. It moves the belt until the first item reaches the end near the operator.

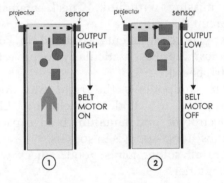

Switches as sensors

Microswitches, magnetic reed switches and tilt switches are often used as sensors.

Things to do

Try to think of an unusual sensing application for a switch or a sensor. For instance, how could we use a photodiode to sense force? Or how could we use a microswitch to sense water level?

Set up a temporary version of your sensor design, and test it. What are its good points? What are its bad points?

The hints in this section are to help you test a project that you have built but does not work. Very often the fault is either a:
- **Short circuit**: an electrical connection that should not be there, or an
- **Open circuit**: a break in a connection.

Checklist

Run through the points in the list below:
- Is the power supply on?
- Is the voltage correct at the positive supply terminal? If it is lower than expected, you probably have a short-circuit somewhere. Switch off immediately.
- Is there a component missing? This includes a missing wire link. Check against the circuit diagram.
- Are all components soldered in correct holes? Applies particularly to projects on stripboard.
- Are components the right way round? Applies to diodes (including LEDs), electrolytic and tantalum bead capacitors, transistors, and ICs in sockets.
- Are all solder joints good? Check with a magnifier.
- Have IC pins become bent when being inserted in their sockets? Remove ICs and check.

Continuity

Check that all points that should be directly connected are in fact connected. Use a multimeter set to its continuity range. A bleep is heard when the two probes are touched to points that are connected. Alternatively, use the LED continuity checker, shown in the drawing at the top of the next column.

Always disconnect the power supply when making continuity checks.

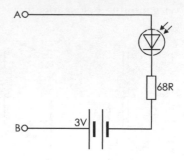

A and B are probes or crocodile clips on the ends of flexible leads about 25 cm long. The LED lights when there is continuity between A and B.

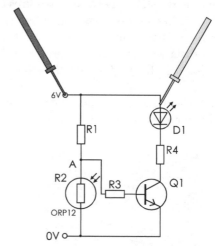

As an example, take the circuit above. The probes should be placed where shown as one of the tests for continuity in the positive supply line. Touch them against the positive supply input terminal and against the anode wire of the LED. Then check between the positive supply input terminal and the terminal wire of R1. In another circuit there might be other direct connections to the positive supply line. These must all be checked for continuity.

In the circuit above, check the negative supply line by placing one probe against the 0V terminal and the other against the terminal wire of R2, and against the emitter terminal wire of Q1.

Short circuits

Short circuts may often cause components to over-heat. There may be a strong smell of scorching, or there may be smoke. If so, turn off the power supply immediately and investigate the cause.

There is a short circiuit if a continuity tester indicates continuity where there should be none. For example, in the circiuit below there should be no continuity across R3. If there is, look for a short circuit. It might be caused by a solder bridge between adjacent tracks.

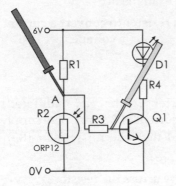

Continuity across a semiconductor device may indicate a faulty device.

Voltage checks

When you are building from a kit, the instructions will often tell you what voltages to expect at certain key points in the circuit. In other cases, it is not difficult to work out what some of the voltages should be.

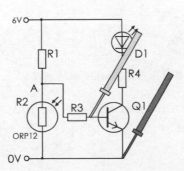

For example, check that the base-emitter voltages of all BJTs are about 0.7 V. If not, the transistor may be faulty. The voltage across the LDR should vary according to the amount of light reaching it. If you place probes on either side of the sensor and wave your hand in front of it, the voltage should rise as the shadow falls on the sensor.

In logic circuits, voltages should be close to 0 V or close to the supply voltage. A voltage that is about half way between usually indicates a short circuit. However, it may also indicate a voltage that is alternating beween high and low levels at a rate too fast for the multimeter to register. If this is suspected, use an oscilloscope to examine the signal. Alternatively, use a logic probe with the facility of responding to rapidly alternating signals.

Signal tracing

Audio and radio-frequency circuits may be tested by injecting a signal (such as a sine wave of suitable frequency) into the circuit at one point, and trying to recover it at another. If the signal is injected at the input to the circuit, it should be possible to trace it at various stages in the circuit until it emerges at the output. It can be traced by using an oscilloscope.

The diagram on the right shows a simple audio signal tracer. If the signal is 'lost' between one stage and the following stage, this helps to locate the fault.

In the reverse procedure, the signal tracer is connected to the output. A signal is injected at the input to the final stage. Then, working *backward* through the stages, the faulty stage can be identified when the signal no longer appears.

27 Interfacing sensors

An **interface** is a connecting link. In this Topic we look at the ways of linking a sensor to the rest of the system. The interface may be just a transistor. Or it may be more complicated.

Transistors

There are several instances of transistor interfacing in earlier Topics. The transistor may be a BJT or an FET. Usually, the transistor is either off, or it is on and saturated.

Signals

Many sensors are resistive. Their resistance changes with temperature, for example, or with light level or position. We use a potential divider to produce a voltage signal from this change of resistance.

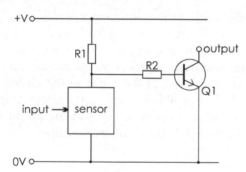

The output is at the collector terminal of the BJT (or at the drain terminal if it is an FET). The output is a varying *current*. Often we connect a load such as a lamp or a relay coil between the output and the positive supply line. There are several examples of this, such as the transistor switch circuits in Topic 23.

In some systems we need a varying **voltage**. Maybe the signal has to be amplified, as in an audio system. In this case we connect a resistor between the collector and the positive supply line. When current flows through the resistor a voltage appears across the resistor. This voltage signal appears at the output.

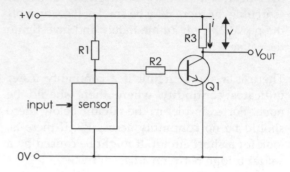

This is an example of:

> **Using a resistor to convert a current into a voltage.**

Example

A transistor (Note: *not* saturated) has a collector current of 3.5 mA. The supply voltage is 9 V and the collector resistor (R3) is 1 kΩ. What is V_{OUT}?

The voltage across the resistor is:

$$v = i \times R3 = 0.0035 \times 1000 = 3.5 \text{ V}$$

One end of the resistor is connected to the positive supply line, so it is at 9 V. If there is a voltage drop of 3.5 V, the other end is at 9 – 3.5 = 5.5 V.

As the current increases, the voltage drop increases and V_{OUT} *falls*. As the current decreases, the voltage drop decreases and V_{OUT} rises. Summing up:

> **The voltage signal is proportional to the inverse of the current signal.**

Self test

1 In the circuit above, h_{fe} = 160, the base current is 30 μA. What is V_{OUT}?

2 If the base current falls to 20 μA, what happens to V_{OUT}?

Two basic variations on this interface circuit are to:

- exchange the sensor and R1, to give the opposite action.
- include a variable resistor in the potential divider to vary the output for a given input

Darlington pair

A Darlington pair consists of two BJTs joined as in the diagram below.

You can wire two single BJTs together or you can buy a ready-made Darlington pair. The ready-made type has the two transistors in a single package with three terminal wires.

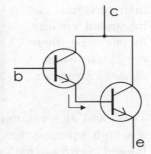

The advantage of a Darlington pair is its gain. This is because the emitter current of one transistor becomes the base current of the other. Assuming that the gain of each transistor is 100, the gain of the pair is 100×100, which equals 10 000. An example of a Darlington is listed in the data table on p. 85.

In operation, there are voltage drops of 0.7 V between the base and emitter of both transistors. This gives a base-emitter voltage of 1.4 V for the pair.

Using a Darlington pair instead of a single BJT greatly increases the input sensitivity of a system.

Schmitt trigger input

When experimenting with some of the transistor switch circuits (examples, p. 69 and 73), you may have thought that the switch-on action is too gradual. As the LDR is shaded, for example, the LED *gradually* brightens. It would be better if it came on suddenly as the light level falls just below the set level. A Schmitt trigger input can produce this effect.

There is another advantage in using a Schmitt trigger. Suppose a system is designed to switch on a porch lamp at dusk. At this time of day, the light level falls very slowly. It may not fall steadily because of clouds occasionally moving in front of the Sun and then clearing again. Or there may be the shadows of leaves moving across the LDR. The effect in a system with a simple transistor switch is to make the porch lamp continually flicker on and off at dusk. This is annoying. Also, if the transistor is switching a relay, the chattering of the contacts will shorten its life.

The performance of the system can be improved by using a Schmitt trigger, as shown below.

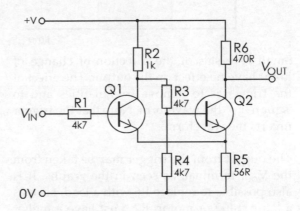

Without going into the theory, the action of the trigger is as shown in the graph below.

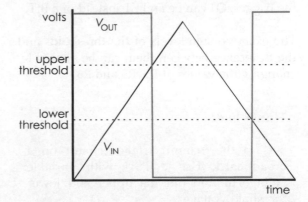

As the input rises from 0 V, the output is first of all high (supply voltage). As the input exceeds the *upper threshold*, the output very rapidly decreases to 0 V.

As the input decreases, the output does not become high again until the input has fallen below the *lower threshold*. Once the input is below the lower threshold, the output does not change again until the input has exceeded the upper threshold.

This graph shows a Schmitt trigger operating on a very irregular input.

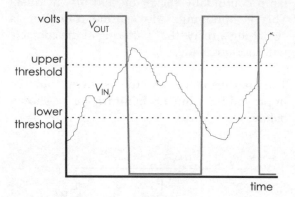

Small reversals of the direction of change of input have no effect on the output. The effect of the trigger is to remove irregularities and to 'square up' the waveform. Note that the trigger inverts the waveform.

The output from the trigger may be taken from the V_{OUT} terminal, as seen in the graphs. It is also possible to replace R6 with a load, such as a lamp, LED or motor. R2 must have a higher resistance than the load. If a large driving current is needed, use a power BJT for Q2. If the sensor can supply only a small current to the trigger, Q1 can be an FET instead of a BJT.

The exact voltage levels of the thresholds and the difference between them can be altered by changing the values of R2, R5 and R6.

Things to do

Set up the Schmitt trigger circuit on a breadboard. Use a PSU with variable voltage to find out the upper and lower threshold voltages.

Experiment with different resistor values to alter the thresholds.

Comparators

A comparator is an amplifying circuit with two inputs. Its output voltage is proportional to the *difference* between the two input voltages. Its gain is about 200 000, so an input difference as small as 100 µV is enough to swing the output well toward 0 V or the supply voltage.

Comparators are manufactured as **integrated circuits**. In one type, there are 23 transistors, 2 diodes, 19 resistors.

These and all the connections between them are built up on a tiny silicon chip. This is housed in an 8-pin package. You do not have to know how many components there are inside the package, or how the comparator works. But you need to know what it does and how to use it.

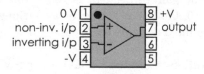

The drawing above shows the 8-pin package. Pin 1 is identified by the circular 'dot' (see photo also). The other pins are numbered as shown.

The two inputs to the amplifier are the non-inverting input (+) and the inverting input (–). The output swings to the positive supply rail if the (+) input is greater than the (–) input. The output swings to 0 V if the (+) input is less than the (–) input.

This device requires a dual power supply. Pin 1 is connected to 0 V. Pin 8 is connected to the positive supply line. Pin 4 is connected to the negative supply line. The positive and negative supplies must be of equal but opposite voltages.

The transistor at the output is like the one in the first diagram on p. 102. It has no collector resistor. At the output, we need to add a resistor connected to a positive supply line.

An example of a complete comparator circuit is:

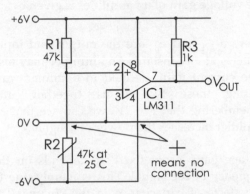

When you read this diagram, note where wires are connected ('dot', see p. 182) and where wires are not connected (no 'dot').

The inverting input (−) is connected to the 0 V supply line, so the inverting input is at 0 V. The non-inverting input (+) is connected to a thermistor voltage divider. If the input from the divider is 0 V or less, the output is 0 V. If the input is a fraction of a millivolt more than 0 V, the output swings very rapidly to + 6V. The output changes from 0 V to +6 V when R2 precisely reaches the set temperature. In this circuit the output changes at about 25°C. Instead of connecting the (−) input to the 0 V line we can connect it to a two-resistor voltage divider that produces a different voltage. The output then changes at a different set temperature.

Design time

Build the touch switch circuit on p. 72, but with a Darlington pair instead of a single transistor. Does this have greater sensitivity?

Then try to drive a heavier load, such as a torch lamp, a motor or a siren. You may need to use a power transistor for the second transistor.

A touch switch with a siren could make a useful panic button in a security system.

A moisture sensor and Schmitt trigger could be the basis of a rain detector. When the washed clothes are out on the line, this could warn someone to bring them in. Design and build a system. Select a suitable warning device.

Design and build a wind detector. Use it as part of a system that flashes an LED when it is windy.

Describe how this circuit works.

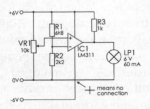

Adapt the moisture sensor to sense when a pot plant needs watering.

Design build and test a system to sound a buzzer when a model train passes a given point on the track.

Draw its system diagram. Suggest an application for this circuit.

28 Amplifying signals

In earlier Topics we have seen how to use a BJT for switching. We can use a BJT on its own, or use two BJTs as a Darlington pair or as a Schmitt trigger. A comparator has a similar switching action. In all these uses, the transistor is used in an all-or-nothing way. It is either off or saturated. All of these are amplifiers because a small change in input results in a large change in output.

Such amplifiers are no use when we need to amplify an audio signal. The waveforms are complicated. When we amplify them, we must maintain their shape as exactly as possible. The aim of the audio amplifier is to produce a varying output voltage that is an exact copy of the input varying voltage, except that the voltages of the output signal are much larger. We turn signal V_{IN} into signal V_{OUT}:

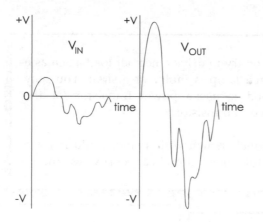

There are other signals that need similar processing. When a doctor is recording an electroencephalogram (EEC), the electrical signals from the muscles of the patients's heart are amplified before being fed to a chart recorder. A seismologist needs accurately amplified signals in order to analyse tremors during an earthquake. At the other end of the frequency scale, we need to be able to amplify ultra-high-frequency signals in a microwave transmitter.

Voltage gain

The **voltage gain** of an amplifier is given by:
$$G_V = v_{OUT}/v_{IN}$$
where v_{OUT} and v_{IN} are the output and input voltages at any instant. An amplifier may also have **current gain**, defined in a similar way. Putting these two gains together, and remembering that $P = IV$, we can see that an amplifier increases the **power** of a signal.

Voltage gain is not fixed. It depends on the frequency of the signal. This is mainly due to the effects of capacitance in the circuits. Try measuring this effect in a low-quality audio amplifier, as below.

Things to do

Connect a signal generator to the input of the amplifier. Connect an oscilloscope to the output of the amplifier.

1 Select an input signal of medium frequency, such as 1 kHz. Select the sine waveform, and a suitable amplitude. Use the oscilloscope to view the amplifier output.

2 Set the amplitude of the input signal to several different values. Measure frequency and amplitude of the corresponding output signals. Calculate the gain of the amplifier. Does the amplifier alter the frequency?

3 Repeat (1) and (2) at lower and higher frequencies, such as 100 Hz, 10 kHz, 100 kHz, 1 MHz and the highest frequency available from the signal generator. Calculate the voltage gain at each frequency.

4 Plot a graph to relate gain to frequency. To plot all the frequencies on one graph, it is best to use a logarithmic scale for frequency, like the one in the next diagram.

A typical amplitude-frequency graph looks like this one, taken on a single-transistor general purpose amplifier:

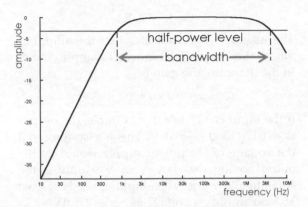

Frequency is plotted on a logarithmic scale, so that each step toward the right represents a tenfold increase in frequency. Amplitude is also plotted on a logarithmic scale. At the top level (0) is the maximum amplitude. Negative values show decreased amplitude. At the level minus 3, the **power** of the amplitude is half the maximum power.

The graph shows that the output signal is below half power at frequencies less than about 800 Hz. It also falls below half power at frequencies greater than about 5 MHz. There is a range of frequencies, from 800 Hz to 5 MHz in which the signal is above its half-power level. This range is defined as the **bandwidth** of the amplifier. In this case, the bandwidth is a little under 5 MHz.

Audio amplifier ICs

Audio amplifiers are widely used in audio equipment such as inexpensive radio receivers, music systems, and intercoms. The amplifiers are manufactured as integrated circuits. They are packaged as 8-pin or 14-pin devices at an average cost of £1. The amplifier is complete except for the addition of a dozen or fewer external components such as resistors and capacitors. The most popular audio amplifiers are the LM380, LM386 and TBA820. For data and constructional details, refer to the data sheets or the suppliers' catalogue.

Operational amplfiers

These integrated circuits, known as **op amps** for short, are widely used for general purpose amplification because of their special features:

- Very high input resistance. They take very little current from the device that supplies them with input.

- Very low output resistance. They are able to supply a large output current without serious drop in output voltage.

- Very high gain.

Like comparators (p. 104), they have two inputs, an inverting and a non-inverting input. Their pin arrangment is slightly different from that of the comparators:

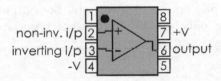

They operate on a dual power supply, +V and –V. They do not need a connection to the 0 V line. However, many op amps can operate on a single supply, in which case, pin 4 goes to the 0 V line.

The output of an op amp swings towards the +V line when the voltage at the (+) input is greater than that at the (–) input. This is the same as for comparators, but op amps do not need a pull-up resistor at the output. The output swings towards the **–V line** when the voltage at the (+) input is less than the (–) input. This is another difference from comparators.

Op amp comparator

Because of their similarity, op amps can be used as comparators. The circuit (overleaf) is similar, except for the power connections and the lack of pull-up resistor.

The output of this circuit swings toward the positive supply as temperature falls:

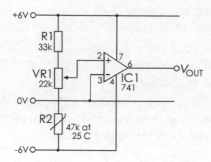

It swings toward the negative supply as temperature rises. The exact temperature at which the outut changes is set by adjusting VR1.

The output of an op amp may not swing all the way to the supply voltages. In the 741, for example, on a ±15 V supply, the output can swing only to ±13 V. In other types it may swing closer.

Inverting amplifier

An inverting amplifier is intended to amplify signals without the output swinging too far in either direction. The connections are like this:

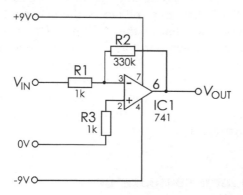

Note that in this diagram (and in some others) the amplifier is drawn with its (–) above the (+) symbol. This is to simplify the layout of the drawing. Always check which terminal is which.

In the inverting amplifier, part of the output signal is fed back to the inverting input. The effect of this **negative feedback** is to reduce the voltage gain to:

$$G_V = -R2/R1$$

The negative sign indicates that the amplifier *inverts* the input signal as well as amplifying it. In the diagram, the gain is:

$$G_V = -330\,000/1000 = -330$$

If the input is +17 mV, for example, the output is $-0.017 \times 330 = -5.61$ V. This is assuming that the voltage of the power supply would let the output swing that far. A ±6 V supply is too small to allow this. The amplifier needs at least ±8 V to amplify a voltage as big as 17 mV.

The resistor R3 is there to compensate for voltage drops across R1 and R2. It has a value equal to R1 and R2 in parallel. Here, R2 is so much larger than R1 that the nearest E24 resistor is 1 kΩ, the same as R1. For lower precision, R3 can be omitted and the (–) input connected directly to the 0 V line.

Self Test

1 An op amp wired as an inverting amplifier has R1 = 2.2 kΩ and R2 = 820 kΩ. What is its voltage gain?

2 Select resistors to give voltage gains of **(a)** –220, **(b)** –12, **(c)** –3, and **(d)** –1200.

Boosting output current

The output of a typical op amp has a resistance of 75 Ω. This means that there is the equivalent of a 75 Ω resistance between the amplifier circuit and the output terminal. If the load is drawing a current of 1 mA, for example, the voltage drop across this resistance of $0.001 \times 75 = 75$ mV. The output of the op amp is 75 mV less than it should be.

If the op amp is to provide more than about 10 mA, its output must be boosted by using a transistor (BJT or FET), as in the circuit opposite.

The load is an 8 Ω loudspeaker. A BC639 transistor has been used, so as to be able to carry up to 1 A.

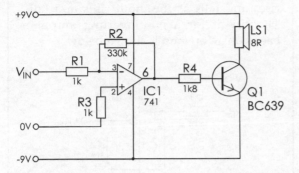

If necessary, a BJT of higher power can be used. For loads that require more than 1 A, the output of the op amp is fed to a Darlington pair (p.103) with a power transistor as the second transistor, or to a power Darlington.

Single supply operation

An op amp is operated on a single supply by using a potential divider to supply the reference line. This is normally the 0 V line, but here the line is at 3 V.

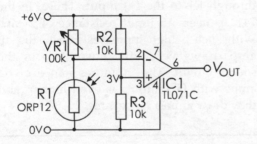

This kind of circuit is useful for battery-powered devices. This one runs on 6 V. However, not all op amps will run on such a low voltage. We have used the TL071C, which runs on a dual ±2 V supply, or on a single 4 V supply. The TL071C has FET inputs, which means that it requires much less input current than the 741, which has BJT inputs.

Another point about the TL071C is that it is a low-noise amplifier.

The output of this circuit is normally at 3 V and swings either side of this. It can not swing fully to 0 V or 6 V.

The LM358 op amp is specially intended for single supply operation, and will run on 3 V (or ±1.5 V). It accepts inputs that are close to the 0V rails and its output will swing down to this level too. It has BJT inputs like the 741, but the input resistance is higher. The LM358 requires less power to operate it than the 741. The LM358 has two independent amplifiers in the same package, which means that amplifying circuits take up less board space. These features make it ideal for battery-powered portable equipment.

Noise

Noise is an unwanted signal that has become added to the signal that is being processed.

Noise comes from various sources, either from inside the circuit or from outside. Noise is generated in several kinds of components, including resistors and transistors. It is generated in components such as amplifiers that contain transistors. In general, the amount of noise generated is related to the size of the currents passing through these components. One way to avoid noise is:

Keep currents as small as possible.

Transistors and op amps can be designed so as to generate   less noise. There are three examples of low noise transistors in the data sheet on p. 85. So another rule for noise reduction is:

Use low-noise versions of transistors and op amps.

Noise that comes from outside the circuit includes voltage spikes on the mains power supply. A refrigerator that is constantly switching its motor on and off produces spikes on the mains supply that affect mains powered circuits in the same house. The spikes may pass through the transformer and into the low-voltage supply. To avoid this:

Place chokes in the power supply circuit.

A similar thing can happen within a circuit where, for example, a relay is continually being switched on and off. Chokes can help. Another technique is to:

Use capacitors for decoupling the supply lines.

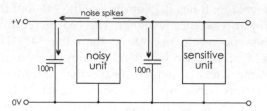

Spikes and pulses from a noisy part of a circuit pass along the supply lines but are absorbed by the capacitors. This is very important in large logic circuits.

Electromagnetic interference (EMI) is caused by magnetic fields from electrical appliances and equipment. Makers of equipment are required by law to keep EMI down to srtrict limits.

EMI is reduced by metal screening, connected to ground or the 0 V line.

This is the reason for grounding the aluminium case of a microphone (p. 97). It is important for signal cable to be screened. Usually the screening consists of a flexible woven sheath of copper wires surrounding the conducting wire or wires. One end of the screening is connected to ground or 0 V line.

This 8-core computer cable is shielded by a braided wire sheath, with an outer layer of insulating plastic.

Non-inverting amp — Box 49

The output of a non-inverting amplifier rises and falls as the input rises and falls. The circuit for a non-inverting amplifier based on an op amp is shown below.

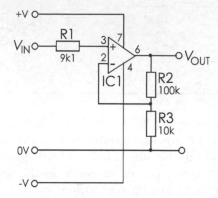

The voltage gain of this amplifier is:

$$G_V = \frac{R1 + R2}{R2}$$

For example, with the values in the diagram:

$$G_V = 110/10 = 11$$

In this amplifier, the input current flows through R1 to the (+) input. There, in the 741, it meets an input resistance of 2 MΩ. With such a high input resistance, the op amp draws very little current from the source circuit. The input resistance of op amps with FET inputs is even higher, and they draw almost no current.

Clipping — Extension Box 50

If the gain of an amplifier is too high, the output voltage can not swing far enough. Then the waveform becomes **clipped**. A clipped audio signal has a distorted sound.

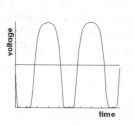

Op amp mixer

An op amp can be used to mix two or more signals. This is a variation on the inverting amplifier. The gain of each signal is:

$$G_V = -R/R_{input}$$

Where R_{input} is the input resistor for the signal. The output is the sum of these amplified signals.

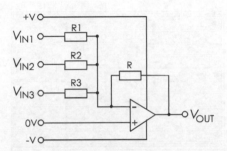

If all resistors are equal, the gain for each is –1. The signals are simply inverted and mixed. If R1, R2 and R3 are unequal, the signals are mixed in different proportions. An audio mixer has variable resistors for R1, R2 and R3, so that the volumes of each channel can be adjusted. A variable resistor for R allows the amplitude of the mixed signal to be increased or decreased.

Self test

An op amp mixer has R1=R2=R3=10 kΩ and R = 22 kΩ. V_{IN1} = 10 mV, V_{IN2} = 15 mV, and V_{IN3} = 5 mV. What is V_{OUT}?

Transistor pre-amplifier

The BJT amplifier below is intended to amplify the signals from a source such as a microphone or photodiode sensor.

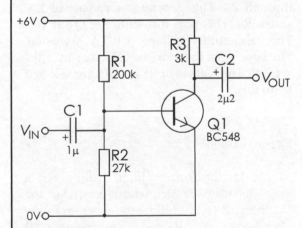

Its features are:
- Resistors R1 and R2 are a voltage divider. They bias the base of Q1 to just over 0.7 V, so the BJT is just switched on but not saturated.
- It may be necessary to alter R1 or R2 slightly depending on the gain of Q1.

- With the biasing resistors as shown and a gain of 400 for Q1, the current through R3 is about 1 mA. This is a small current to minimise noise.
- A current of 1 mA through R3 causes a voltage drop of 3 V. This puts the voltage at the collector about half way between 0 V and the positive supply. This allows the collector voltage to swing freely up or down.
- C1 is a **coupling capacitor** that passes the signal from the microphone to the base of Q1. Its value is selected so that it will pass all frequencies in the audio range (30 Hz to 20 kHz).
- C2 is another coupling capacitor that passes the amplified signal at the collector to a further amplifying stage.
- The varying base current produces a larger and varying collector current. This produces a varying voltage across R3. The collector voltage is an amplified and inverted signal. The gain of this amplifier is about –100.

More about this and other amplifiers overleaf.

Building and testing the pre-amplifier   Extension Box 53

Things to do

1 Build the pre-amplifier (p. 111) either on a breadboard or on stripboard (see right and p. 74).

2 Measure the voltages at the base and collector of Q1 to check that they are suitable. If not, replace R2 or possibly R1.

3 As input, use a signal generator, set to give a sine wave at 1 kHz and with amplitude 25 mV. Observe the output signal with an oscilloscope. Measure its amplitude. What is the voltage gain of this amplifier?

4 Check that the gain of the amplifier is constant over the whole audio range.

The diagram below shows how to lay out the components on a piece of stripboard. This is a top view. The black dots at A1, E1 and J1 are 1 mm terminal pins. The diagram shows the copper strips, but these are *under* the board. You can see where to cut one of the strips at E8. A wire link runs from G13 to J13.

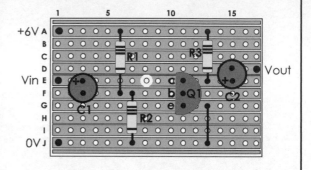

A stabilised pre-amplifier   Extension Box 54

The additon of an emitter resistor (R4) with a large-value capacitor (C3) reduces the gain of the amplifier but improves its stability. It makes the gain of the amplifier independent of the gain of the transistor. Because the gain of the transistor depends on temperature, the modified circuit is less dependent on temperature. It is stabilised.

When the 'no signal' collector current of 1 mA is flowing through R3, an approximately equal current is flowing through R4. This generates a voltage of 1 V across R4. This puts the emitter of Q1 at 1 V. The base-emitter voltage is 0.7 V, as usual. The base must therefore be biased at 1.0 + 0.7 = 1.7 V. The values of R1 and R2 are selected to do this.

Things to do

1 Build the preamplifier, either on a breadboard or on stripboard (p. 74). For the stripboard layout follow the diagram in the box above. Add R4 and C3 in the space left vacant to bottom right. Do not include the wire link from G13 to J13.

2 Test the amplifier as described in the previous box.

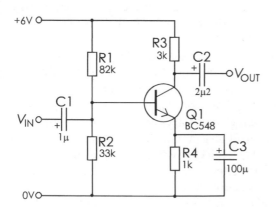

Op amp inverting amplifier

This circuit uses almost any op amp, that has a single amplifier in the package. Instead of a double supply, with 0V line and positive and negative supply lines, the op amp has a single supply (0 V and +V), with R2 and R3 forming a voltage divider. The divider provides a +3 V reference line.

Things to do

Build the amplifier on a breadboard or on stripboard (p. 75).

1 Select a suitable op amp from data sheets.

2 Decide what gain you require and select the appropriate resistors.

3 Test it as described opposite

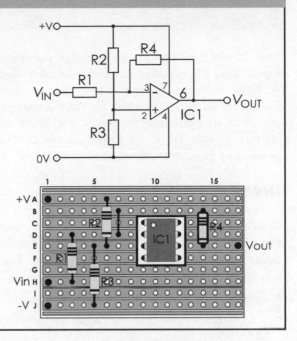

Questions on systems, sensors and interfacing

1 Draw a system diagram of a electronic system that you might find in an office. Identify the three main stages of the sytem.

2 Describe a strain gauge, how it works and one application in which a strain gauge is used.

3 Describe how would obtain an amplified signal from a named type of microphone.

4 What is a Hall effect device? Describe an application in which a Hall effect device is used. Draw a system diagram to explain how it is used in this application.

5 Describe a simple moisture sensor and how to use it to switch a relay.

6 Why are position sensors inportant in industrial machinery. Describe three position sensors used in industrial applications.

7 What is a Darlington pair? How does it work? If the voltage gains of the transistors are 120 and 220, what is the gain of the pair?

8 Describe the action of a Schmitt trigger? Draw a diagram to show how it is used.

Questions on amplifiers

9 What is meant by the bandwidth of an amplifier?

10 List the main features of an operational amplify and explain why they are important.

11 Draw the circuit of an op amp comparator circuit. How would you use the op amp as a comparator to process the signal from a thermistor sensor?

12 Describe how the output of an op amp comparator differs from that of a regular comparator IC (such as a 311).

13 Draw a circuit diagram of an op amp inverting amplifier. How do we calculate its voltage gain? Give an example.

14 Show how to operate an op amp on a single supply.

15 What is noise? What are its causes? How may noise be avoided?

Extension question

16 Describe the circuit of an op amp non-inverting amplifier.

29 Timing

The circuits that we have looked at in previous topics have all operated instantly. At least, this is how it seems to us. They take a few nanoseconds to respond but, in effect, a change of input immediately produces a change of output. In this Topic we introduce noticeable periods of time into circuit operation.

Delay

Charging a capacitor takes time (p. 43). The time taken to charge a capacitor introduces a delay into the operation of a circuit. The sequence is:

- Discharge the capacitor completely.

- Let current flow into the capacitor through a resistor.

- Wait until the voltage across the capacitor reaches a set level.

- The delay is the time taken to charge to the set level.

The problem is to monitor the voltage across the capacitor. We need to do this without drawing currrent from the capacitor. The solution is to use an op amp which has *high-resistance* inputs.

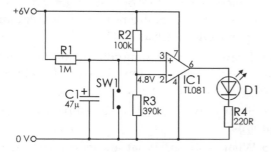

In the circuit above, current flows through R1 and charges C1. We press SW1 to discharge C1 at the start of the delay.

R2 and R3 form a voltage divider. The voltage at the (–) input of IC1 is 4.8 V.

When S1 is pressed briefly, the voltage across C1 and at the (+) input becomes zero. IC1 is acting as a comparator (p. 107). As C1 charges, the voltage at (+) rises. At first, it is less than the voltage at the (–) input, so the output of IC1 is close to 0 V. The LED is not lit.

When the voltage at (+) reaches 4.8 V and above, it is greater than the voltage at (–). The output of IC1 swings up to nearly 6 V. The LED is lit. The time taken for charging C1 from 0 V to 4.8 V is approximately 75 s. When S1 is pressed, the LED goes out and there is a delay of 75 s before it comes on again.

This circuit works because IC1 has:

- **High input resistance**. It takes very little current from C1.

- **Low output resistance**. It can provide plenty of current to light the LED.

When a device is being used in this way, we say that it is a **buffer** between the capacitor part of the circuit and the LED part of the circuit.

Things to do

Set up the delay circuit on a breadboard.

1 Measure the delay.

2 Use a digital multimeter to measure the voltage across C1 as it charges. Why is an analogue meter of no use for this measurement?

3 How can you increase or decrease the delay time?

4 Draw a system diagram of the delay circuit.

5 Re-design the delay circuit to switch a lamp *on* when the button is pressed. This could be used to light a lamp in a dark corridor for a period, then switch it off automatically. Think of other applications.

Pulse generator

Instead of thinking of the previous circuit as a delay, we can think of it as a pulse generator. It produces a low pulse for 75 s, which turns off the LED for that period of time.

Here is a circuit for a pulse generator (or delay) that is based on two BJTs.

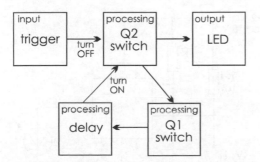

The diagram shows a possible output to an LED that indicates the state of the circuit.

The circuit consists of two transistor switches. The output from each switch is the input to the other. A system diagram shows that the connection from Q2 to Q1 is direct (through R3), but the connection from Q1 to Q2 passes through a delay stage. The delay stage is provided by C1 and R2 — another example of charging a capacitor through a resistor.

The circuit is triggered by briefly connecting the trigger input to 0 V. Triggering the circuit turns Q2 off. This puts the LED on (remember, transistor switches are inverting switches).

Without the delay, Q1 would turn Q2 on again immediately. You would not see even a flash of light from the LED. With the delay unit present, there is a delay of a few seconds while C1 charges through R2. Q2 is turned on when C1 has charged to the right level.

This circuit is stable when Q1 is off and Q2 is on. It remains indefinitely in that state.

It is unstable in the reverse state, with Q1 on and Q2 off. After the delay, it goes back to the other state. A circuit that is stable in only one of two states is called a **monostable**. There are several kinds of monostable. They are used to generate a single pulse when triggered.

Triggering the monostable

The monostable is triggered by briefly bringing the base of Q2 to 0 V to turn it off. Instead of doing this by hand we can use a sensor. Here is one way of interfacing the monostable to an LDR.

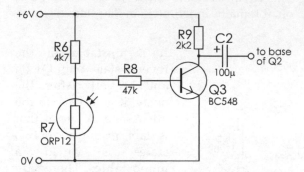

The interface is a transistor switch that takes its input from a voltage divider R6/R7. The switch turns on when a shadow falls across R7. This causes a fall in voltage at the collector of Q3. This is passed across the coupling capacitor C2 to the base of Q2 in the monostable circuit. We do not need the diode D1 when using this interface.

This is a good example of using a capacitor to couple two circuits together. Normally, the collector of Q3 and one side of the capacitor are at 6 V. Normally, the base of Q2 is at 0.7 V. The capacitor has a voltage difference of 5.3 V between its two plates. When a shadow falls on R7 the voltage at the collector drops suddenly to zero, a fall of 6 V. The voltage on the other plate of C2 falls by an equal amount, from 0.7 V to –5.3 V, maintaining for a while the 5.3 V difference across C3. The fall of voltage at the base of Q2 turns it off, triggering the monostable action. The system diagram is:

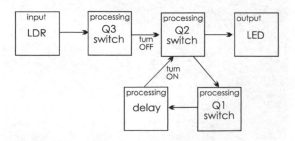

The 555 timer IC

This IC can be used as the basis of a monostable circuit. The advantages of the 555 are:

- Greater precision in the length of the pulse.

- Pulse length is not affected by variations in the supply voltage.

- A long pulse length (up to 1 hour) is obtainable.

- Only a few additional components are needed.

- Output current is up to 200 mA, which is enough to light a lamp or power a relay.

A low-power version of the 555 is known as the 7555. This can generate even longer pulses, with output current up to 100 mA.

Below is the standard monostable circuit for the 7555.

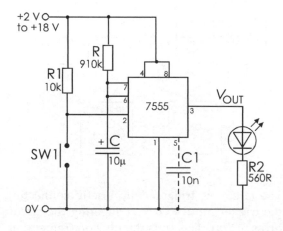

The 555 operates on a supply of 4.5 V to 16 V. It requires the capacitor C1 connected between pin 5 and the 0 V line. This capacitor is not needed with the 7555.

The length of the pulse depends on the timing resistor R and the timing capacitor C. The length, t, of the pulse is:

$$t = 1.1RC$$

t is in seconds, R is in ohms and C is in farads. With the values in the diagram, the pulse lasts for 10 s.

The trigger input at pin 2 is normally held at the positive supply voltage. In the diagram we show a pull-up resistor R1 that does this. The timer is triggered by a brief low pulse. Here we have a push-button that, when pressed briefly, connects the trigger input to the 0 V line. The input can be triggered in other ways by interfacing various sensors to the timer.

The output (pin 3) is normally at 0 V. It rises instantly to the supply voltage when the timer in triggered. It falls instantly to 0 V at the end of the pulse.

Pin 4 is the reset pin. Often this is permanently wired to the positive supply line, as in the diagram opposite. If this pin is briefly connected to 0 V while the timer is generating a pulse, the output of the timer at once goes to 0 V.

Things to do

Connect a resistor and push-button to reset the timer.

Astable circuits

This kind of circuit is like a monostable that triggers itself to start again at the end of pulse. The result is a circuit that is *not* stable (astable) in any state. It runs continuously, producing pulses indefinitely.

A clear example of this is the circuit below. This is similar to the BJT monostable (p. 115) but with a delay circuit for each transistor switch.

Things to do

Build and test the monostable circuit, using a 555 or 7555 timer IC.

1 Try replacing R with resistors of other values and see what effect this has on the pulse length.

2 Try replacing C with capacitors of other values and see what effect this has on the pulse length.

3 Using the equation given above, try to build timers with pulse lengths of 20 s, 60 s, 5 min, and 0.1 s. Use an oscilloscope to check the 0.1 s pulse.

4 Interface the timer to an LDR sensor. Find values for R and C so that the LED lights for 30 s when a shadow falls on the LDR.

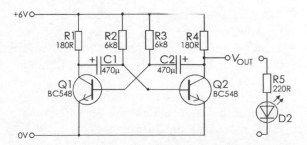

The system diagram of the astable is:

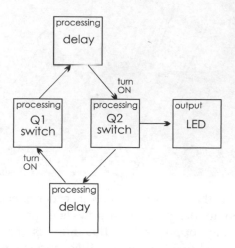

This is a system without a special input stage. The astable runs for as long as the power is switched on. However, the power switch (not shown in the drawing) can be thought of as an input. Another possible input is a switch that closes to connect the base of Q2 to the 0 V line. This holds Q2 off and prevents the astable from running. It runs again as soon as the switch is opened.

Things to do

Build the astable on a breadboard.

1 How long are the pulses? How long are the gaps (LED off) between the pulses? Do these timings agree with times calculated from the equations above?

2 Try the effect of replacing C1 and C2 with capacitors of other values, such as 47μF. If the LED flashes very quickly, use an oscilloscope to measure the frequency.

3 Try the effect of replacing R2 and R3 with resistors of other values, such as 680Ω.

4 Investigate what happens when C1 and C2 are not equal in value, or when R2 and R3 are not equal.

5 Try to get the astable running at several hundred hertz. Replace the LED with some other more suitable output stage.

The 555 astable

The 555 or 7555 timer IC can be used to build an astable circuit:

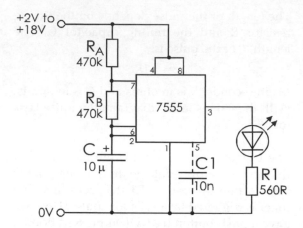

In the astable, the timing capacitor C is charged by current flowing from the positive supply through the timing resistors R_A and R_B. While this is happening, the output at pin 3 is equal to the supply voltage. The time taken for C to charge depends on the values of R_A, R_B and C.

As soon as the capacitor is charged, it is immediately discharged through R_B and pin 7. While this is happening, the output at pin 3 is at 0 V. The time taken depends on the values of R_B and C, but not on R_A.

One end of C is connected to the trigger input (pin 2, see p. 117). As C fully discharges, the falling voltage at pin 2 triggers the IC to start charging C again. With the values shown in the diagram, the IC takes about 10 s to run through a complete charge/discharge cycle.

The time, t, for one complete cycle is:

$$t = \frac{(R_A + 2 \times R_B)C}{1.44}$$

The time, t_1, for which the output is high is:
$$t_1 = 0.69(R_A + R_B)C$$
The time, t_2, for which the output is low is:

$$t_2 = 0.69R_BC$$

It is clear from the equations that t_1 is greater than t_2. We can make them *almost* equal by making R_A small compared with R_B. If we must have the periods exactly equal, or if t_1 must be smaller than t_2, we use a circuit with diodes, as in Extension Box 56.

For heavier loads, the output of a 555 or 7555, either in monostable or astable mode, can be used to drive a transistor switch. The circuit is the same as is used for boosting the output of an op amp (p. 108-9).

Things to do

Build and test a 555 or 7555 astable.

Duty cycle — Extension Box 56

The duty cycle of an astable is:

$$\text{Duty cycle} = \frac{t_1}{t_1 + t_2} \times 100\%$$

A 555 astable circuit such as that opposite always has a duty cycle greater than 50%. With a circuit like the one below, the lengths of t_1 and t_2 can be set independently.

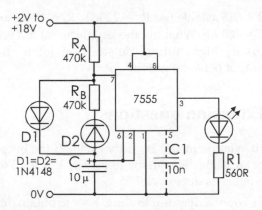

The capacitor is charged through R_A and D1. It is discharged through D2 and R_B.

$$t_1 = 0.69 R_A C$$
$$t_2 = 0.69 R_B C$$

Variable frequency — Box 57

To obtain variable frequency, make R_A small (say, 1 kΩ) so that the duty cycle is always close to 50%. For R_B, use a variable resistor with a small fixed resistor in series with it. The fixed resistor (say, 470 Ω) is there to prevent a zero resistance between pins 6 and 7 when the wiper of the variable resistor is at one end of its track.

Variable duty cycle (1) — Box 58

The duty cycle can be varied by replacing resistors R_A and R_B by a variable resistor. The wiper is connected to pin 7. In this diagram, R_A is replaced by R1 and the 'top' part of VR1. R_B is replaced by the 'bottom' half of VR1 and R2.

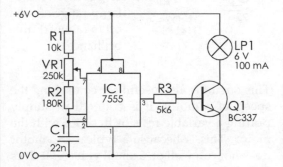

The duty cycle can be adjusted from about 70% to almost 100%. The astable runs at about 170 Hz. The lamp is turned on and off at this rate. It flickers too rapidly for the eye to see. It appears to be on continuously.

This is a lamp dimming circuit. The brightness can be varied smoothly over the range of 70% to 100%.

When the duty cycle is less than 100%, the lamp is being supplied with power for only part of the time. The shorter the 'on' period and the longer the 'off' period, the dimmer the lamp.

Variable duty cycle (2) Box 59

The disadvantage of the circuit in Extension Box 58 is that the duty cycle must always be 50% or more. This means that it is not possible to dim out the lamp completely. To give full dimout, we need to be able to reduce the duty cycle to almost zero. We use diodes as in the circuit in Extension Box 56.

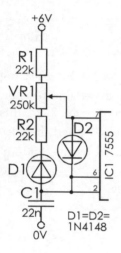

R1 and the 'top' part of VR1 represent R_A. The 'bottom' part of VR1 and R2 represent R_B. Now the duty cycle can range from 3.7% to 96.3%. The brightness of the lamp ranges from off to (almost) full brilliance.

This circuit is also useful for controlling the speed of a motor. It is better than simply putting a variable resistor in series with the motor. This is because a typical DC motor does not run well on a small voltage. It does not start well and, when running has a tendency to stall. This is particularly likely to happen if the mechanical load on the motor is suddenly increased.

Using the 7555 astable control, variation in the duty cycle causes variation in the *length* of the pulses, not variations in their voltage. The turning action of the motor is always at full strength. The result is that the motor can be run slowly without stalling.

Provided that the astable runs fast enough (a few hundred Hertz), there is no jerkiness in the action of the motor.

Questions on timing

1 Describe a delay circuit built from a resistor, a capacitor and an op amp. Why is the op amp needed?

2 Draw the circuit diagram of a monostable based on two BJTs.

3 Draw the system diagram of a BJT monostable.

4 What are the advantages of a monostable circuit based on a 555 or 7555 timer IC?

5 With the help of a circuit diagram, describe a practical application for the 555 or 7555 IC when used as a monostable.

6 Given the values of the timing resistor and timing capacitor, calculate the pulse lengths produced by a 555 monostable:

(a) R = 47 kΩ and C = 100 nF.

(b) R = 10 kΩ and C = 2.2 nF.

(c) R = 2.2 MΩ and C = 470 µF.

7 Given that the timing capacitor of a 555 monostable is 39 nF, what is the nearest E24 value of resistor required to produce a pulse of:

(a) 1.15 ms.

(b) 5 µs.

(c) 22 ms.

8 With the aid of a circuit diagram and system diagram, describe the action of a BJT astable.

9 A 555 astable has R_A = 22 kΩ, R_B = 47 kΩ and C = 100 nF. What are the lengths of (a) its cycle, (b) its high output pulses, and (c) its low output between pulses?

Extension questions

10 What is meant by duty cycle? If the output of an astable is high for 45 ms and low for 5 ms, what is its duty cycle?

11 Draw a diagram to show how to obtain duty cycles of 50% or less from a 555 astable.

12 Describe a circuit that uses a 555 astable as a motor speed control. What is its advantage over a variable resistor in series with the motor?

Design time

This page has some ideas for projects based on timing. Design and build!

A BJT monostable is used at a 'pulse stretcher' in an electronic version of the twisty wire game. When the loop touches the wire, the monostable is triggered for about 5 seconds, ringing a bell or flashing a lamp for that time.

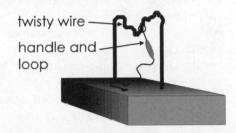

twisty wire

handle and loop

A porch lamp or a lamp for a dark corridor is to be switched on for 2 minutes by pressing a button briefly. Use a 555 timer in monostable mode.

An egg timer is based on a 555 monostable that switches on an LED. Press a button when you put the egg in the boiling water. The LED comes on until the egg is ready, 5 minutes later.

Try to improve the design by adding a switch to select cooking periods of 4, 5 and 6 minutes.

A simple buzzer or single-note siren (p. 159) is cheap. But the continuous sound it gives out is easy to ignore. Build an astable (either BJT or 555) that makes the siren 'beep' about once a second. This makes the sound more attention-catching.

A metronome circuit flashes an LED at a rate that can be set to several different tempi, from *largo* to *presto*. Preferably make the duty cycle low, so that the LED flashes crisply on the beat.

A 555 astable can be the basis for an audio signal generator. You can drive the speaker with a transistor switch. The frequency of the generator should be adjustable over the range 50 Hz to 1 kHz.

Finally, a rather tricky one. The panic button, when pressed, makes this project emit a note at about 1 kHz for a period of 3 minutes.

To make it even trickier, let the note sound for half-second 'bleeps' with a half-second gap between them.

There are lots of other timing projects that you could design. Think up some for yourself!

30 Logic

Logic circuits are used for processing **binary** information. By 'binary', we mean that the information has only two possible states. For example, a switch is open or it is closed. It can not be half-open or half-closed.

There are two switches in the circuit below. There is one lamp. The circuit has two binary inputs and one output.

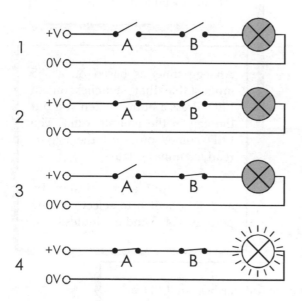

There are four possible ways in which the two switches can be set:
1 A open and B open: lamp off.
2 A closed but B open: lamp off.
3 A open but B closed: lamp off.
4 A closed AND B closed: lamp ON.

There is only one way to light the lamp — close A AND B.

The circuit performs a **logical operation.**, the **AND** operation. The circuit works only for closing the switches. Leaving A and B open has no distinctive result. Note the binary nature of the inputs. The switches are open or closed. Note the binary nature of the output. The lamp is on or off.

This circuit can have practical applications. The switches could be microswitches that detect the positions of two safety grids on a power drill. The switches close when the grids are locked in position. The lamp comes on when *both* grids (A AND B) are in position. This signals 'OK to drill' to the operator.

The action of the circuit can be summarised if we represent the binary states of inputs and output by '0' and '1'.
For the switches, 0 = 'switch open' and 1 = 'switch closed'.
For the lamp, 0 = 'lamp off' and 1 = 'lamp on'.

Now we can set out the four states of the switches in a **truth table**:

| Inputs | | Output |
B	A	Z
0	0	0
0	1	0
1	0	0
1	1	1

The table shows the lamp on only when both A AND B are closed.

Logic is not limited to two inputs. There may be any number. For example, there could be four switches in series. The lamp lights only if A AND B AND C AND D are all closed.

There are instances when a few switches wired together can perform logic, as in the example of the drill given above. However, there is a limit to what can be done with mechanical switches. The next example shows how we use electronic logic circuits for the purpose. They are faster and cheaper, and we can build complex logic functions in a very small space.

Electronic logic

Electronic logic circuits work with two levels of voltage:

- **Low:** 0 V or close to 0 V.
- **High:** The positive supply voltage, or close to it. In some types of logic circuit, 'high' is always 5 V. In others, it may have other values.

Usually the low voltage level corresponds to logical '0' and the high level to logical '1'.

To see how logic circuits operate, we will consider a practical example of a logical system. This is part of a security system that controls a floodlight in the garden of a house. Intruders are detected when they break a beam of infrared light that is directed at a photodiode. The floodlight is to be switched on when the beam is broken. However, there is no point in turning on the floodlight during the daytime, so there is an LDR set to tell whether it is day or night.

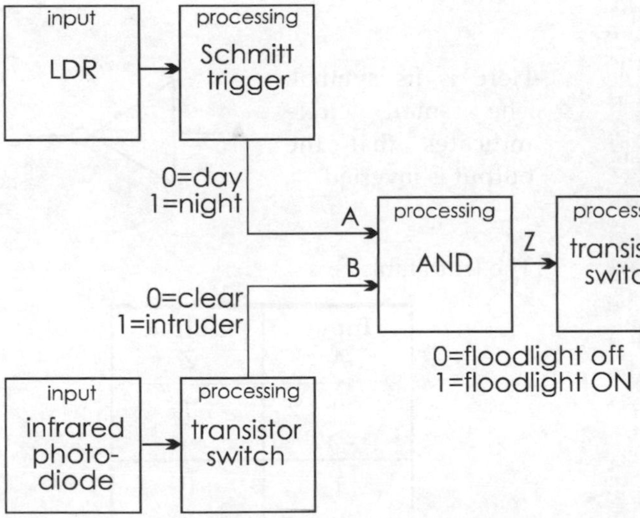

The system diagram shows the two sensors and the processing of their outputs by two interface circuits. These are a Schmitt trigger and a transistor switch. The diagram shows which logic levels from the interfaces correspond with which conditions at the inputs.

The logic signals from the interfaces both go to the same next stage. This is a logic circuit that performs the AND operation. It works according to the truth table opposite. It accepts two inputs, A and B, and produces a single output, Z. According to the table, the floodlights are switched on only when it is night AND an intruder is detected.

Logic gates

Building logic circuits is simple. All the logic gates and other more complicated logic circuits that you might need are available as integrated circuits. There are two commonly used 'families' of logic ICs:

- **TTL**, which is short for transistor-transistor logic. This runs on 5 V, so it needs a regulated power supply. TTL type numbers all begin with '74' so this is sometimes known as the 74XX series. There are various types of TTL, of which the Low Power Schottky type has almost replaced the original 74XX series. 74LSXX ICs need less power than the 74XX type.

- **CMOS**, which is short for complementary MOS. These have type numbers ranging upward from 4000, so are sometimes known as the '4000' series. They run on any voltage between 3 V and 15 V.

CMOS is slower than TTL but requires less current. It has the additional advantage that it does not require a regulated power supply.

Many of the 74XX series are also available as CMOS ICs. Their type numbers begin 74HC. They operate on 2 V to 6 V, require less current than TTL and are faster than CMOS.

Logic ICs

Both TTL and CMOS are packaged as double-in-line ICs, usually with 14 or 16 pins, sometimes more. There are four AND gates to an IC, sharing the power supply pins.

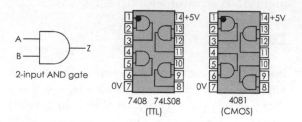

2-input AND gate

7408 74LS08
(TTL)

4081
(CMOS)

The connections to the gates are different in the two types of IC.

Design tip

The two types of logic IC differ in the way they must be used.

1 TTL must have a 5 V regulated supply. CMOS can run from batteries.

2 All inputs of a CMOS IC must be connected to 0 V, the positive supply, a CMOS output, or another point in the circuit. If you leave any input unconnected while testing a circuit, the IC will not work properly. It may overheat and destroy itself.

3 TTL inputs may be left unconnected. An unconnected input behaves as if it has a high (='1') signal applied to it. For best operation, the unused inputs should be connected to the positive supply through a 1 kΩ resistor. Several inputs can share the resistor.

OR gate

This performs another common logical operation. In words, the output of an OR gate is high (1) if any one OR more of the inputs is high.

On the right is the symbol for a 2-input OR gate. Any larger number of inputs is possible.

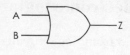

The truth table is:

Inputs		Output
B	**A**	**Z**
0	0	0
0	1	1
1	0	1
1	1	1

OR gates are available in 14-pin ICs, with the same pinouts as shown on the left.

NOT gate

This gate is unusual because it has only one input. The output is always the inverse of the inputs. This is why it is also called the INVERT gate.

Here is its symbol. The small circle indicates that the output is inverted.

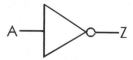

The truth table is:

Input A	Output Z
0	1
1	0

Systems of gates

The output of a gate can be fed to the input of one or more other gates. In this way we can create more complex logic functions.

As an example, take this system of two AND gates:

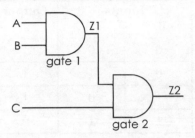

There are eight possible combinations of the three inputs and this is their truth table:

| Inputs | | | Z1 | Output |
C	B	A		Z2
0	0	0	0	0
0	0	1	0	0
0	1	0	0	0
0	1	1	1	0
1	0	0	0	0
1	0	1	0	0
1	1	0	0	0
1	1	1	1	1

The stages of working out the table are:

1 Draw a blank table with the headings given above.

2 Fill in the eight combinations of inputs.

3 Work out the logic at gate 1. Looking only at inputs A and B, work down the column Z1. Enter the value of A AND B for each combination of inputs. Use the truth table on p. 122 to help with this.

4 Now we can go on to work out the logic at gate 2. Looking only at columns C and Z1, work down column Z2. Enter the value of C AND Z1 or each combination of inputs to gate 2.

5 Z2 is the output of the system. This is '1' only when all three inputs are '1'. In other words, these two gates are the equivalent of a 3-input AND gate.

Here is another example:

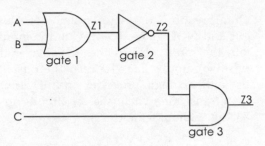

As before, there are eight possible combinations of inputs:

| Inputs | | | Z1 | Z2 | O/P |
C	B	A			Z3
0	0	0	0	1	0
0	0	1	1	0	0
0	1	0	1	0	0
0	1	1	1	0	0
1	0	0	0	1	1
1	0	1	1	0	0
1	1	0	1	0	0
1	1	1	1	0	0

The stages of working out the table are:

1 Draw a blank table with the headings given above.

2 Fill in the eight combinations of inputs.

3 Work out the logic at gate 1. Looking only at inputs A and B, work down the column Z1. Enter the value of A OR B for each combination of inputs.

4 In Z2 enter the inverts of the values in Z1 (0 for 1, and 1 for 0).

5 Looking only at columns C and Z2, work down column Z3. Enter the value of C AND Z2 or each combination of inputs to gate 3.

6 Z3 is the output of the system. This is '1' only when A and B are '0' and C is '1'. There is no single gate equivalent to this system.

Things to do

On a breadboard, set up systems of AND, OR and NOT gates. To provide the inputs, join the input pins to the 0 V line or the positive supply line. To read the outputs, use either a logic probe, or an LED driven by a transistor switch (see the circuits on p. 73, but without R1 and R2).

For each system of gates, run though all possible combinations of inputs. Fill in a truth table to show how the output varies with combinations of inputs.

Self test

Write out the truth table of each of the systems of gates shown below. Where possible, draw the equivalent single gate that could replace them.

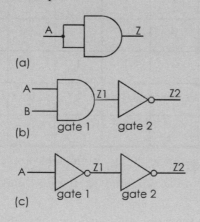

(a)

(b)

(c)

NAND gate

The NAND gate or NOT-AND gate is one of the most useful gates for logical processing (see Extension Box 66). The NAND gate is the equivalent of an AND gate followed by NOT, as in diagram (b) above.

The symbol is an AND gate with a small circle at its output to indicate that the output is inverted.

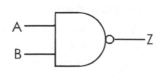

The output is '0' only when A AND B are '1', as shown in the truth table:

| Inputs | | Output |
B	A	Z
0	0	1
0	1	1
1	0	1
1	1	0

Two-input NAND gates are available as ICs, with four gates to the package.

NOR gate

This is the equivalent of an OR gate followed by a NOT gate. Its truth table is:

| Inputs | | Output |
B	A	Z
0	0	1
0	1	0
1	0	0
1	1	0

The output is '0' when A OR B are '1'.

The symbol is an OR gate with a small circle at its output to indicate that the output is inverted.

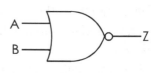

Self test

Write the truth tables of these circuits:

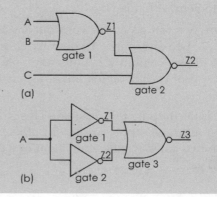

(a)

(b)

Ex-OR gate

An Ex-OR, or exclusive-OR gate is a relative of the OR gate. The output of an OR gate is '1' when inputs A OR B **OR BOTH** are '1'. The output of an Ex-OR gate is '1' when inputs A OR B but **NOT BOTH** are high. The truth table differs from that of OR in the last line:

Inputs		Output
B	A	Z
0	0	0
0	1	1
1	0	1
1	1	0

This is sometimes known as the 'same or different' gate. Output is '0' if the inputs are the same and '1' if they are different. This can be useful in circuits when we want to compare two logical quantities to see if they are alike or not.

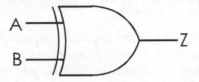

The ex-OR gate has only 2 inputs.

An **ex-NOR gate** is an ex-OR followed by NOT. Its output in the inverse of ex-OR.

Logical switching

Obtaining logical AND with mechanical switches is described on p. 122. By connecting the switches in parallel we obtain logical OR. The lamp lights when any one or more of the switches is closed.

A system such as this provides control over a device from one of a number of stations. An example is a fire alarm system in which the alarm is switched on by closing any one of a number of switches located in different parts of a building.

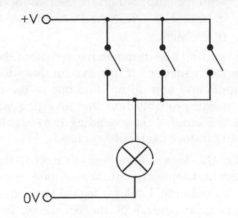

Boolean notation

On previous pages we have represented logical operators by writing out the name in full. For example, the action of a two-input AND gate is written:

$$Z = A \text{ AND } B$$

This can also be written using a 'dot' (.) for 'AND'. The equation is written:

$$Z = A.B$$

The OR operation is represented by a 'plus' symbol. Thus, the action of a three-input OR gate is:

$$Z = A + B + C$$

The NOT or INVERT operation is indicated by placing a bar or line over the variable. So the action of a NOT gate is written:

$$Z = \overline{A}$$

Logic circuit design 1

Follow this plan:

- Set out a truth table with a row for each combination of input states. With one input there are two states (0 and 1). With two inputs there are four combinations of states (00, 01, 10, and 11). With three inputs there are eight combinations (000 to 111).

- Fill in a column for each output. On each row enter '0' or '1', according to what you want to happen to that output for each combination of inputs.

- Taking each output column in turn, scan down it and look for one of these:

  * An exact match with one of the input columns.

  * An exact inverse of one of the input columns.

  * A set of '0's and '1's that is identical with the output of one of the two-input gates: AND, OR, NAND, NOR ex-OR or ex-NOR.

- If you find one of the above, write out the logical equation. If the system has two inputs and you do not find one of these, you will probably find that inverting *one* of the inputs before sending it to a gate will produce the result you need.

- Use the logical equations to design the system, keeping the number of gates used to a minimum. Look for logical terms that appear in several of the equations. For example, you may find that A occurs in two or more of the equations. If so, you need to invert A only once and use it in obtaining several outputs.

Half adder

A half adder circuit takes two one-digit binary numbers and adds them together. The are only two different one-digit binary numbers — '0' and '1'. These are to be added in all four possible combinations.

The first step in designing the logic is to set out the truth table. This describes the inputs and what output we expect them to produce. Here is the table for a half adder:

| Inputs | | Outputs | |
B	A	Carry	Sum
0	0	0	0
0	1	0	1
1	0	0	1
1	1	1	0

This is ordinary binary arithmetic, in which:

$$0 + 0 = 0$$
$$0 + 1 = 1$$
$$1 + 0 = 1$$
$$1 + 1 = 0, \text{carry } 1$$

Looking down the C column, we can see that C is '1' only when both A and B are '1'. This is an example of the AND operation. As a logic equation, we write:

$$C = A.B$$

The column for S is recognisable as the output of an exclusive-OR gate. The Boolean equation is:

$$S = A \oplus B$$

The symbol $\oplus$ means 'exclusive-OR'.

The logic circuit for these two equations is:

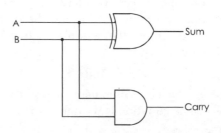

Logic circuit design 2

Sprinkler control

A system is designed to turn on a garden water sprinkler when the soil is dry, but not when the sun is shining. A light sensor A has outputs 0 = dull, and 1 = sunny. A soil moisture sensor B has outputs 0 = moist, and 1 = dry. For the sprinkler S, 0 = off, and 1 = on. The truth table for the system is:

Inputs		Output
B	A	S
0	0	0
0	1	0
1	0	1
1	1	0

The sequence of '0's and '1's in the output column does not fit any of the suggestions in the plan opposite. The third line down is 'dull day and dry soil', the ideal conditions for watering. For S = 1, we see that A = 0 and B = 1. As a logic equation, this is written:

$$S = \overline{A}.B$$

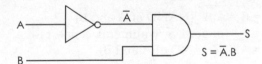

This is a case of inverting one of the inputs to a standard 2-input gate.

Chaser lights display

A fairground booth has a row of lamps that are switched on in pairs, the pattern travelling from left to right.

direction of travel ⟶

All lamps labelled with the same letter are switched on together.

The system needs only four outputs, W, X, Y, and Z. So it needs only two inputs. A circuit that is described in Topic 31 produces the inputs 00, 01, 10 and 11 in a repeating sequence. The truth table is:

Inputs		Outputs			
B	A	W	X	Y	Z
0	0	1	1	0	0
0	1	0	1	1	0
1	0	0	0	1	1
1	1	1	0	0	1

1 = 'lamp on' and this travels across the output side of the table from left to right.

Scanning the columns:
- Z is the same as B:
$$Z = B$$

- X is the inverse of B:
$$X = \overline{B}$$

- W has the same outputs as an ex-NOR gate:
$$W = A \oplus B$$

- Y has the same outputs as an ex-OR gate:
$$Y = A \oplus B$$

The logic circuit is:

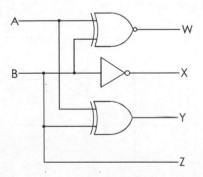

This circuit produces the required result. However, it might be possible to build it more economically, as explained in Extension Box 66.

Logic circuit design 3

Traffic lights

The sequence of lighting the red, yellow and green lamps goes through four stages, as shown in the truth table:

Inputs		Outputs		
B	A	R	Y	G
0	0	1	0	0
0	1	1	1	0
1	0	0	0	1
1	1	0	1	0

For four stages, we need two inputs A and B. The output columns conform to the standard cycle. Scanning the columns, we note that:

- Y is the same as A, so:

$$Y = A$$

- R is the inverse of B, so:

$$R = \overline{B}$$

- With three '0's and one '1', G looks like the output of an AND gate, but the bottom two lines are in the wrong order. Invert the A input, and obtain:

$$G = \overline{A}.B$$

The logic circuit is:

Decoder

You have probably noticed that the inputs to a 2-input system in the truth table are the binary numbers 00, 01, 10, and 11. These are equivalent to decimal numbers 0, 1, 2, and 3.

This system decodes the binary numbers to drive a 7-segment display (p. 158). Each of the segments a to g has an output to drive it.

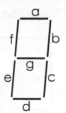

The truth table is:

Inputs		Outputs						
B	A	a	b	c	d	e	f	g
0	0	1	1	1	1	1	1	0
0	1	0	1	1	0	0	0	0
1	0	1	1	0	1	1	0	1
1	1	1	1	1	1	0	0	1

Logic equations are:

$a = d = \overline{A.\overline{B}}$ (NAND of A and $\overline{B}$)
$b = 1$ (on for every numeral, connect to +V)
$c = \overline{\overline{A}.B}$ (NAND of $\overline{A}$ and B)
$e = \overline{A}$
$f = \overline{A + B}$
$g = B$

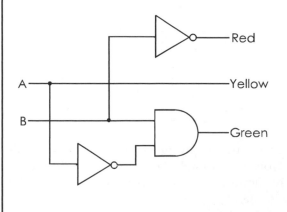

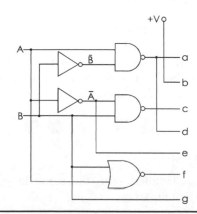

NAND logic

Extension Box 66

NAND gates can be wired so that they are equivalent to any other kind of gate.

The diagrams on the right show how to connect NAND gates so that they have the same action as the other kinds of gate.

NOT
A
$\overline{A}$
AND
A
B
$\overline{A.B}$
A.B
OR
A
B
$\overline{A}$
$\overline{B}$
A + B
NOR
A
B
$\overline{A}$
$\overline{B}$
A + B
$\overline{A + B}$
ex-OR
A
B
$\overline{A.B}$
A⊕B
ex-NOR
A
B
$\overline{A.B}$
$\overline{A⊕B}$

Things to do

Work out the truth tables of the diagrams and confirm that they perform as stated.

Alternatively, using ICs on a breadboard or a ready-made logic board, connect up some of these NAND gates circuits, and discover the truth tables.

The reason for making other gates out of NAND gates is a matter of economy and of simplifying the layout of circuit boards.

Logic gates are manufactured as ICs, each of which contains four two-input gates (p. 124). NOT gates are six to a package. To be most economical of cost, as many as possible of the gates should be used. Look at the traffic-light circuit opposite This has two NOT gates and an AND gate. Two different kinds of gate means that we have to have *two* ICs. Using two NOT gates leaves four gates unused (but needing current). Using only one AND gate leaves three unused. When converted to an all-NAND equivalent (right), it needs only four NANDs, all in *one* IC.

By converting the circuit we have saved a whole IC, an IC socket and the board space that they occupy. Also the wiring is much simplified by having only one IC to which to make connections.

Red
NOT
A
Yellow
Green
NOT
B
AND

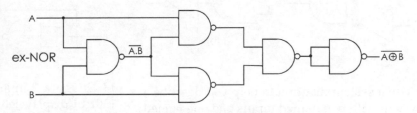

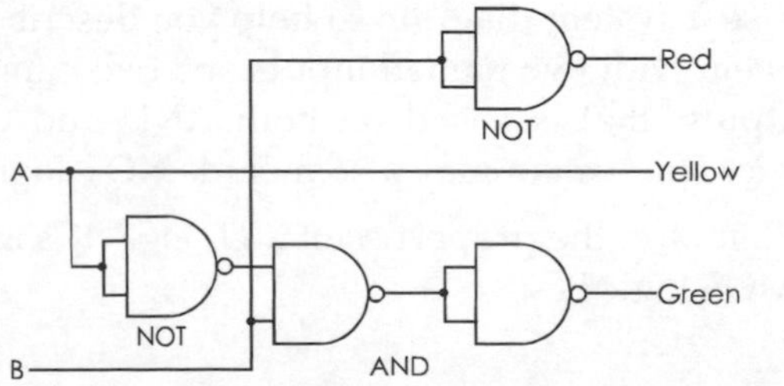

Questions on logic

Use this combined truth table for reference when answering these quesions and for the 'Design Time' circuits opposite.

| Inputs | | Outputs | | | | | |
B	A	A.B	$\overline{A.B}$	A+B	$\overline{A+B}$	A⊕B	$\overline{A⊕B}$
0	0	0	1	0	1	0	1
0	1	0	1	1	0	1	0
1	0	0	1	1	0	1	0
1	1	1	0	1	0	0	1

1 What is meant by 'binary'? Give some examples to illustrate this idea.

2 Describe a switched logic circuit to control the electric motor of a drilling machine and make the machine safer to use.

3 Write the truth tables of these logic circuits:

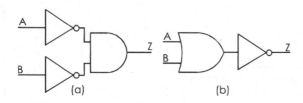
(a)   (b)

4 Use a system diagram to help you describe a system with two named inputs and one named output, that is based on AND logic.

5 Use a system diagram to help you describe a system with three named inputs and one named output, that is based on OR logic. The system may also include NOT logic.

6 Use a system diagram to help you describe a system with two named inputs and two named outputs, that is based on both AND and OR logic. The system may also include NOT logic.

7 Compare the properties of TTL logic ICs and CMOS logic ICs.

8 A system is designed to switch on a room heater if the temperature is below 15°C, but not if the door has been left open. Draw the system diagram, including the logic required.

9 A system is designed to flash a beacon lamp and sound a siren when any one of three windows or the door is opened. But the beacon is not to be flashed at night. Draw the system diagram, including the logic required.

10 Write the truth table for the logic circuit below. To what gate is it equivalent?

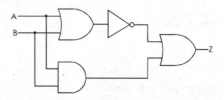

Extension questions

11 Design a logic circuit for the chaser lights display (p. 129) using only NAND gates.

12 Write the truth table for this circuit. To what gate is it equivalent?

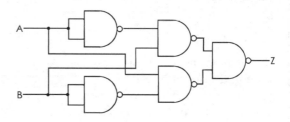

13 Design a 2-input decoder circuit (as on p. 130) to display letters a to d on a 7-segment display, as below:

14 For the circuit below, write the value of Z as a Boolean expression, using A and B.

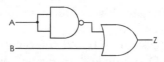

15 Draw the circuit that has output $Z = \overline{A + B + \overline{C}}$.

Design time

Build the chaser light display circuit (p. 129) and use it to drive 12 LEDs arranged in a circle.

Build the sprinkler control system (p. 129). Design your own soil moisture sensor. Use a solenoid-operated valve to turn on the water, or a small fountain pump driven by an electric motor.

A model railway is on display in a shop window. The train runs continuously in the direction shown by the arrows. There are two sensors (A and B) that detect when the train is passing. For A and B, 0 = no train, 1 = train passing. The train is to circulate the loops alternately.

The junction can switch left or right. Its motor responds to two control signals, Z1 and Z2. For Z1, 0 = motor off, 1 = motor on. For Z2, 0 = go left, 1 = go right.

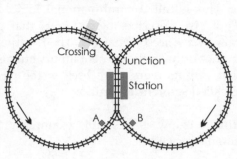

Draw a truth table of the operation of the junction. Note that one combination of inputs A and B is not possible. Design the logic circuit.

As extensions of the system, add logic and other circuits to (a) make the train halt in the station for 1 minute every time it goes through and (b) stop the train if it approaches the crossing and the gates are closed across the track.

Build the decoder system (p. 130). Use logic that produces counts 0 to 3 (as on p. 136). Or you could try out the logic for your solution to Q. 13 opposite.

Alternatively, redesign the logic to give 1, 2, 3, and 4, or to give capitals A, B, C and D.

Feed the output of a slow-running 555 astable to a NOT gate. What do you notice about the duty cycle of the output from the gate? A useful tip!

In an industrial process, two liquids are run into a tank through valves W and X and stirred together by stirrer Y. Then the mixture is run out of the tank through valve Z. There are two level sensors A and B which have output 0 when they are uncovered and 1 when they are covered. Starting with a full tank (A covered) and the stirrer running, valves W and X are to be closed and valve Z is to be opened. When the mixture has drained to that both sensors are uncovered, valve Z is to close and valves W and X are to open. The stirrer continues running. The tank then fills until both sensors are covered. The cycle repeats indefinitely.

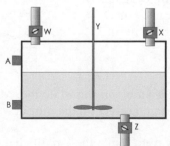

Design a logic system to produce the action specified above. Then add a manual over-ride switch that immediately closes W and X, and opens Z. The stirrer stops when the lower sensor is uncovered and the mixture is allowed to drain away completely.

31 Logical sequences

The logic gates described in Topic 30 produce a fixed output for every combination of inputs. Each gate has a truth table that describes its action. We can connect several gates together in a logical circuit and write a truth table that describes the outputs for every *combination* of inputs. This is called **combinational logic**. In this Topic we describe logic circuits that go through a *sequence* of changes. Their outputs depend not only on the present inputs but also on what inputs there have been in the past. This is called **sequential logic**.

The circuit below is a simple example of a sequential logic circuit.

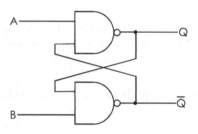

The output of one gate connects back to an input of the other gate. This is a reminder of the circuits on p.115 and p. 117. These circuits have two possible states. This circuit too has two possible states. It is stable in both states, so it is called a **bistable** circuit.

If you try working out the circuit using the NAND truth table, or if you try some practical runs on an actual IC, you find that the circuit is stable only when both inputs are high (=1). When it is stable, one of its outputs is low (=0) and the other is high (=1). In one of its stable states the logic levels are like this:

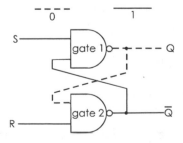

The inputs have been relabelled S (= set) and R (= reset). In the drawing, the circuit is in the reset state, with its Q output low (0). Its $\overline{Q}$ output is the reverse of this. The bar over the $\overline{Q}$ indicates this relationship.

In the drawing, the inputs appear not to be connected to anything. In practice, they must be connected to something that will hold them at a high logic level. A pull-up resistor connected to the positive supply line would do.

Watch what happens if we make the S input low for an instant:

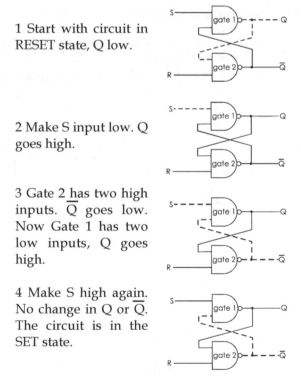

1 Start with circuit in RESET state, Q low.

2 Make S input low. Q goes high.

3 Gate 2 has two high inputs. $\overline{Q}$ goes low. Now Gate 1 has two low inputs, Q goes high.

4 Make S high again. No change in Q or $\overline{Q}$. The circuit is in the SET state.

The circuit is stable in this state too. Making S low *again* has **no effect** on the output.

The only way to make the circuit change back to its reset state is to make the R input low for an instant. The resulting action is the reverse of that above.

S-R bistables

The circuit described oppposite is a **set-reset bistable**. It is also known as a set-reset **flip-flop**.

The result of a brief low pulse to one of its inputs may be 'no change' or a change of state from reset to set, or from set to reset. It all depends on the state the circuit is already in. It depends on what has happened to the circuit in the past.. In this way, the circuit has a **memory**. Set-reset bistables are used as the units of certain types of digital memory.

Self test

S-R bistables can also be built from a pair of NOR gates. Draw the diagram and work out the logic levels as the circuit is set and reset.

Design tip

S-R bistables are a useful circuit unit because they can be set or reset by the output from sensors or from other logic. Also, they are easy to build using a pair of gates from a NAND or NOR IC. The next example show you how to use a NAND bistable to 'remember' that there is an intruder around.

Using a S-R bistable.

On pages 69 and 73 are circuits that switch on a lamp or LED whenever the light falling on an LDR is reduced. These circuits could be used with an electronic siren instead of LP1 or D1.

This could be the beginning of a security system. A beam of light shines on an LDR. If an intruder passes through the beam, the siren sounds. This is not quite good enough, for the siren stops sounding as soon as the intruder moves out of the beam. The system needs to 'remember' that the beam has been broken. This is where we use a bistable.

The system diagram looks like this:

The circuit might be:

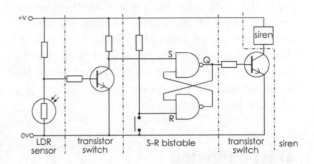

When the beam is broken the transistor switch is turned on. It delivers a low pulse to the S-R bistable.This makes the output (Q) go high. This turns on the second transistor switch, which switches on the siren.

Once the beam is broken, the siren sounds continuously until someone (a) switches off the power supply, or (b) presses the reset push-button. Pressing the button produces a short low pulse at the reset input. The bistable changes state. Q goes low and the siren is switched off.

Clocked logic

In a big logic system there may be dozens, hundreds, or even thousands of gates. Any of them may change state when they receive a suitable input. Different gates take different lengths of time to change. The situation is very complicated. It is like an orchestra with all the players following their scores at their own speed. It needs a conductor. Watching the conductor's baton, the players all keep in step and the music sounds right.

The conductor of a logic system is the **system clock**. This might be a 555 astable circuit.

The system clock (or simply 'clock', as we shall call it) produces a series of pulses at a fixed rate. In very large systems, such as a computer, the clock beats very fast. It may run at several hundred megahertz.

In a clocked logic system, the logic circuits do not act immediately there is a change in their inputs. Instead, they all wait. They do nothing until the clock tells them to act. Most act on the 'rising edge' of the clock. That is, the instant when the signal from the clock changes from low to high. However, some logic is clocked by a falling edge, so you must always check the data sheets for this point when designing a clocked logic circuit.

D-type flip-flop

This is our first example of a clocked logic device. The device actually consists of several logic gates joined together, but we will just look at the action of the circuit *as a whole*. The circuit is available as a CMOS integrated circuit (4013). Each IC contains two identical flip-flops. The drawing shows the symbol for a D-type flip-flop.

There are four input terminals:

- D, the data input.
- CLK, the clock input.
- Set and reset.

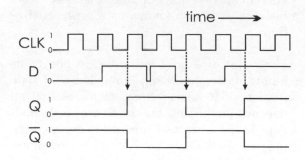

There are two output terminals, Q and $\overline{Q}$. $\overline{Q}$ is the inverse of Q.

Data is fed to the D input. The data may be high or low and may be changing, but nothing happens to the outputs. Then, when the clock input rises from low to high, the data that is present at the D terminal *at that instant* appears at the Q output. If the input changes again, there is no change at Q. There are no further changes at Q until the *next* rising edge of the clock signal. At all times, $\overline{Q}$ is the inverse of Q.

The graph below plots typical changes in the inputs and outputs of a D-type flip-flop.

Although D changes, there is no change in Q until the next rising edge. Then Q changes to be equal to D at that time. Note that in the sequence above the second time D changes, it changes back again *before* the next rising edge of the clock. In this case, the change in D does not register as a change in Q.

A D-type flip-flop acts as a **latch**. It samples the data input at regular intervals. The value is held until the next sampling. This is useful if data is changing rapidly. It gives a logic circuit time to process data without it changing while it is being processed.

The Set and Reset inputs are used when we need to change the output immediately, without waiting for the clock. Normally these two inputs are held low and have no effect. If the Set input is made high, Q immediately goes high and $\overline{Q}$ goes low. The reverse happens if the Reset input is made high.

Counter/divider circuit

The D-type flip-flop can be built into several useful circuits.

The Set and Reset inputs are not used. They are connected to the 0 V line. The $\overline{Q}$ output is connected to the D input.

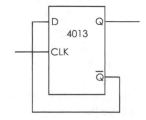

The stages of its operation are as follows:

1 Suppose the Q output is 0. The $\overline{Q}$ output is 1, and this is fed back to the D input.

2 At the next rising edge of the clock, Q takes the value of D and becomes 1. $\overline{Q}$ changes to the inverse, 0, and this is fed back to D.

3 At the next rising clock edge, Q becomes 0. $\overline{Q}$ changes to 1, and this is fed back to D. The circuit is now back to stage 1 and the cycle repeats indefinitely.

Look at this as a graph:

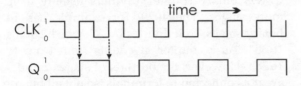

Q changes every *other* time that the clock changes. In other words, it changes at half the rate of the clock. Or, we can say that the frequency of Q is half the clock frequency. The flip-flop is acting as a **frequency divider**.

Design tip

If the output of an astable is fed to the clock input of this circuit, the output Q of the circuit has half the frequency of the astable. The more important fact is that Q has a duty cycle of *exactly* 50%, no matter what the duty cycle of the astable. This is a simple way of obtaining a 50% duty cycle.

Two-stage divider

If the Q output from the 4013 is fed to the CLK input of a second 4013 (perhaps the other flip-flop in the IC), we divide the frequency again. The output of the second flip-flop has a frequency that is one quarter of the original frequency. By setting up a chain of flip-flops, we obtain half-frequency, quarter-frequency, eighth-frequency, and so on.

An interesting action comes from feeding the second flip-flop with the $\overline{Q}$ output.

The circuit is:

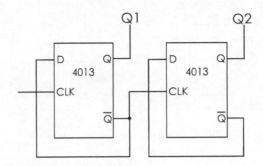

The graph of the outputs is:

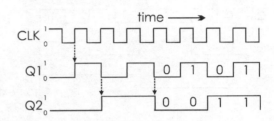

The frequency is halved at each stage, as before. However, the logic levels of Q1 and Q2 go through an interesting sequence. In the table below, we set them out more clearly:

Q2	Q1	Binary	Decimal
0	0	00	0
0	1	01	1
1	0	10	2
1	1	11	3
0	0	00	0
0	1	01	1
And so on ...			

Beacuse the CLK input for Q2 comes from the inverted output of Q1, Q2 changes on the *falling* edge of Q1. Taking the Q1 and Q2 logic levels as numbers written in binary, we see that the output from the flip-flops repeatedly runs through the decimal equivalents: 0, 1, 2, 3, 0, 1, ... and so on. The circuit is a **counter**.

We can add more stages to the counter

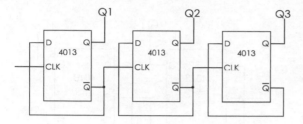

The output graphs are:

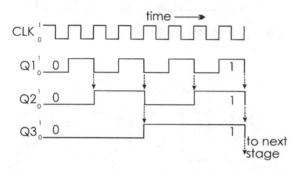

The falling edge of each stage triggers the next stage to change state. This produces the binary equivalents of 0 to 7, repeating. If we add a fourth flip-flop to the chain, the counter runs from 0000 to 1111, repeating. This is equivalent to 0 to 15 in decimal.

The graphs show that the counter circuit also acts as a frequency divider. It divides the frequency by two at each stage.

Other binary counter/dividers

A four-stage binary counter/divider is available as a single IC, the 74LS93. The counting chain is in two parts, one containing a single flip-flop and the other containing three flip-flops already connected as in the drawing on the left. This gives three dividing options: divide by 2, divide by 8 and (by connecting the two parts with an external link) divide by 16. In terms of counting, the OIC can count from 0 up to 1, 7, or 15.

CMOS binary counter/dividers usually have more stages. The 4020, for example has 14 binary stages, so it divides by 2^{14}, which equals 16 384. As a counter, it counts from 0 up to 16 383. However, the outputs of stages 2 and 3 are not connected to terminals, so not all binary numbers can be obtained.

The 4040 IC has 12 stages, dividing by 2^{12}, or 4096. It counts from 0 to 4095. Outputs are available from all stages. The 4060 IC has 14 stages, like the 4020, though it does not have outputs from stages 1 to 3 and 11. Its advantage is that is has an astable circuit included. It needs only a capacitor and a pair of resistors to build the clock generator. For further details of these ICs, refer to the data sheets.

Things to do

Build a two-stage counter/divider, using the two flip-flops of a 4013 IC. The diagram below shows how to connect the power supply and the Set and Reset terminals.

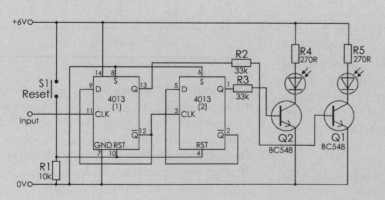

Take the input from a slow 555 or 7555 astable or a signal generator. Observe the counting action as indicated by the LEDs.

Extend the counter by adding two more stages and two more LEDs.

Decade counters

In Topic 32, several ICs are identified by type number so that you know which ones to use when experimenting, or for building a project. The aim is to give you an idea of the range of counters that is available. However, apart from the 4013 and (later) the 4017, you are not expected to memorise type numbers and their distinctive features. All the details you may need are provided in data sheets or in retailers' catalogues.

Although digital circuits work most easily with binary numbers, humans are more familiar with decimal numbers. Several types of counter IC produce decimal output. The output runs through the sequence from 0000 to 1001 (0 to 9) and then returns to 0000 to repeat the sequence. The 74LS90 is an example of a decade counter IC.

Another decade counter is the 4518, which has two decade counters in one IC (see 'Cascading counters', below). The 4510 contains a counter that has the useful feature of being able to count up or down. If its up/down input is low, it counts up in the normal way (0, 1, 2, ... , 9, 0, 1, ...). If this input is high, it counts down (9, 8, 7, ..., 1, 0, 9, ...). Another feature of this counter is that it has four inputs that are used to load the counter with any starting value from 0 to 9. Probably the most versatile of the decade counters is the 4029, which has all the features of the 4510, and may also be set to count either in binary or in decimal.

Cascading counters

A single decade counter runs from 0 to 9. If we want to count numbers larger than 9, we need to join two or more counters in series. This is called **cascading** them. The 4518 is cascaded as shown in the diagram at top right.

Some counters have special outputs for use when cascading. The 4518 does not have such an output so we use an AND gate to detect when the output of the counter is '9', by ANDing the '1' and '8' outputs.

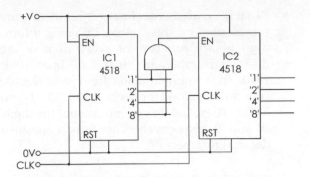

Note that the clock input drives *both* counters. The logic levels are:

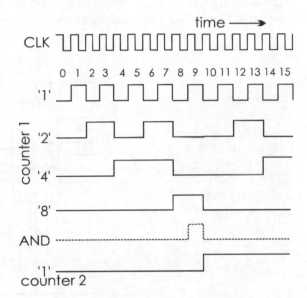

There are several points to note in this graph:

- **Decade counting:** The 4518 counts from 0 (0000) up to 9 (1001), then goes back to 0 and starts again. This gives a sequence of outputs different from binary counting (see opposite page).

- **Carrying:** A counter operates only while its ENABLE input is high. Counter 1 has this tied to the +V line It counts all the time. The EN input of Counter 2 is fed from the AND gate. The output of this is normally low, so Counter 2 ignores the clock pulses. However, when Counter 1 gets to the count of '9', the output of the AND gate goes high. With a high EN input, Counter 2 advances on the next clock pulse. Then the EN input goes low and Counter 2 does not count again until after the next '9' on Counter 1.

139

- **Binary coded decimal:** Numbers are represented on this counter by using a four-digit binary number for every digit of the decimal number. In binary, decimal 10 is 1010. In a counter made from two decade counters, the code for decimal 10 is '0001' and '0000'. These binary codes represent the digits '1' and '0' respectively. They thus represent a decimal number.

As another example of binary coded decimal (or **BCD**), take the decimal number 25. In binary, this is '11001'. In BCD, it is '0010' and '0101', representing '2' and '5' respectively. BCD is important for driving numeric displays.

The 4511 has four inputs for the binary values '1', '2', '4', and '8'. It has seven outputs for the seven segments of a numeric display. These outputs are different from the usual CMOS outputs because they provide enough current (up to 25 mA each) to drive the LED segments directly without using transistor switches.

The current from each output passes through a resistor that limits it to a safe amount. Each LED of the display has its own anode terminal but the cathodes are connected within the display to a common terminal. This type of display is called a **common cathode** display.

Below is a typical circuit for using the 4511:

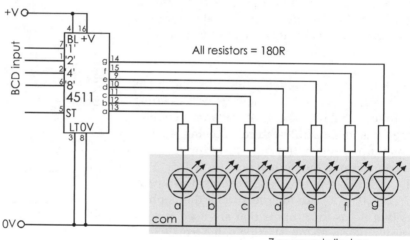

7-segment display

BCD decoders

On p. 130 we described a circuit that decodes the first four binary numbers to produce the logic levels for driving a seven-segment display. There are ICs that contain decoding circuits for the whole range of binary numbers from 0 to 9. The 4033 is a decade counter with a built-in seven-segment decoder. Instead of the usual '1', '2', '4', and '8' outputs, it has seven outputs that will drive a display. The outputs are not able to drive the LEDs directly, so you need seven transistor switches for that purpose. There are also decoders that can take their input from sources other than counters. One of the most popular of these is the 4511.

The 4511 has latches that store the input. When the STORE input is low, the output follows the input. If the STORE input is made high, the input data is held latched. The output shows the latched data, in decoded form. Any further changes that occur in the data arriving at the IC have no effect on the display. This is useful because it makes it possible to hold a rapidly changing input, to allow time to read the display.

The BLANK input is normally held high. If it is made low, the LEDs all go out. This happens also if the IC receives an input corresponding to numbers 10 to 15. The LAMP TEST input is normally low. If it is made high, all segments come on. This allows the display to be tested.

Resetting the counter

Sometimes we want a counter that counts up to a number other than 10 or 16. For example, we might want to count up to 6, then continue from zero. This is easily arranged by external gates. The technique is to generate a pulse that resets the counter as soon as it reaches the number *after* the required maximum count. Here is a circuit for making a 4518 count up to 6:

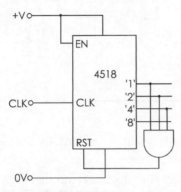

The counter counts normally from 0 to 6. During these stages there is no stage in which all inputs to the AND gate are high. Its output stays low. On the next clock pulse, the outputs are the binary equivalent of '7'. Outputs '1', '2', and '4' all go high. Immediately, the 3-input AND gate detects there are three highs and its output goes high. The high level at the RESET input resets the counter to 0000. There is in fact a brief count of 7, but resetting is too fast for this to be noticed.

Design tip

Unless you are going to use several 3-input AND gates for other purposes in your circuit, you would probably not use a 3-input AND gate for resetting. Instead, you could use a 3-input NAND gate and follow this with an inverter. The inverter too might be another NAND gate with its inputs joined together (p. 131).

Any surplus 3-input NAND gates can easily be converted to 2-input NAND gates by joining two of their inputs together.

The 4017 counter

The output stage of this counter is completely different from that of other counters looked at so far. It is a decade counter and its outputs are described as 'one of 10' outputs. The IC contains a special decoder to provide these outputs.

It has 10 output terminals all of which, except for the '0' output, go low when the counter is reset. Then, on each rising edge of the clock input, one of the outputs goes high, in order from 1 to 9.

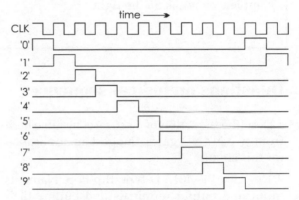

This type of counter is useful for triggering a series of actions. As each output goes high, it triggers an action. An example is driving a display of changing coloured lights. A complex sequence of lighting effects can easily be produced. If any stages are to go on for longer than the others, they may be triggered repeatedly by consecutive outputs fed to an OR or NOR gate.

It is simple to arrange for this counter to produce a lower count. If we want it to count to 7, for example, the output '8' is fed back to the RESET input. The counter has an OUT output. This goes high for counts 0 to 4 and low for counts 5 to 9. Two 4017 counters may be cascaded by feeding this output to the CLOCK input of the second counter.

The 4017 is useful for dividing a frequency by 10 or, if cascaded, for dividing by 100 or higher powers of 10. Note that cascading does not extend the 'one of 10' action to '1 of 20' or more.

Two-digit counter

The single-digit 0 to 9 counter may be extended to make a two-digit 00 to 99 counter.

The first stage consists of two decade counters with a gate to decode the output of Counter 1. This enables Counter 2 on the count of '9'. The circuit for this is on p. 139.

The second stage consists of two 7-segment displays. These are each connected as in the diagram on p. 140. A system diagram of the complete circuit is on the right.

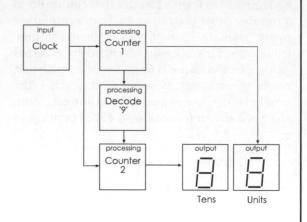

Questions on logical sequences

1 Describe the action of a set-reset bistable built from two NAND gates. Suggest an application for this circuit.

2 Describe the 4013 D-type flip-flop. Name its input and output terminals and outline their functions.

3 Draw a graph to show the changes in input and output of a D-type flip-flop as the logic levels of the data change. What is the advantage of a system that is based on clocked logic?

4 Describe how a D-type flip-flop is used as a data latch.

5 Draw a circuit in which a D-type flip-flop is used to divide the frequency of a signal by two. Draw a graph of all input and output signals.

6 Draw a circuit in which two D-type flip-flops are used to build a two-stage counter. Draw a graph of all input and output signals.

7 Describe the features of three different types of counter IC and suggest a possible application for each type. Type numbers are not required.

8 Describe how two decade counter ICs are cascaded to count from 0 to 99.

9 Explain what is meant by 'binary coded decimal'.

10 Write the decimal number '69' (a) as a binary number, and (b) in BCD.

11 What is meant when we say that an IC decodes BCD into levels for a 7-segment display?

12 Taking any digit between '0' and '9', write it in BCD form and in the form in which it is used to drive a 7-segment display.

13 Describe a circuit that can be used to make a decade counter count up to 4 instead of up to 9.

14 Describe a circuit that can be used to make a 4-stage binary counter count up to 12 instead of up to 15.

15 Describe the 4017 decade counter IC. Draw a graph of its inputs and outputs while it counts from 0 to 4. Suggest an application for this counter IC.

Extension questions

17 Using a system diagram, illustrate how to build a 2-digit display circuit that counts from 00 to 99.

18 In what way could the circuit of Q. 17 be modified to count from 00 to 45?

Design time

More circuits for you to design and (possibly) build.

An alarm sounds briefly when someone stands on a pressure mat. The alarm continues to sound after the person has stepped off the mat. It stops only when a reset button is pressed.

The system sounds an alarm, as on the left, but the sound continues for 5 minutes after the person has stepped off the mat, then stops automatically.

A metronome circuit with variable speed flashes a red LED on every beat. It also flashes a green LED on every fourth beat.

Design and build a system that counts people passing through a doorway and displays the number.

Next, try to modify the system so that it rings a bell when the 7th person passes through the doorway.

A clock running at 50 Hz or more drives a D-type flip-flop. The outputs of this switch on two LEDs alternately. One LED is labelled 'Heads', the other 'Tails'. There is a button which, when pressed, stops the clock. This leaves either 'Heads' or 'Tails lit, virtually at random.

A simple reaction tester has a 4017 IC and a row of 10 LEDs which light up one at a time, in order. The counter is driven by a 10 Hz astable. Design the logic so that the counter starts when the operator presses a button. The subject watches for the first LED to light and presses a button as soon as they see it. Pressing this button stops the counter. The subject's reaction time to the nearest tenth of a second is measured by noting which LED is lit.

Increase the frequency of the astable to obtain greater precision.

An alarm sounder unit that emits an intermittent (beep-beep-beep-...) sound, instead of a continuous sound.

Slightly more complicated is a unit that emits a two-toned sound.

32 Storing data

The processing stages of some systems may need to store data. The data, or **information**, is stored in the system to be used later.

The set-reset bistable (p. 134) is an example of a circuit that stores data.

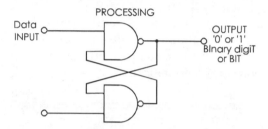

It 'remembers' if the Set input has been made low since the bistable was last reset. Its output is a low voltage or a high one. We may write its output as '0' (usually for 'low') or '1'. These are the two digits used for numbers in the binary system. The output of the bistable is a single **BI**nary digi**T**, or **BIT**.

The D-type latch (p. 136) is another memory device with a one-bit output.

Bits and bytes

A bit can have one of two possible values, 0 or 1. The bit can represent facts, for example, 0 = 'It is not raining', and 1 = 'It is raining'. The facts represented by '0' and '1' must always be the exact opposites of each other. Alternatively a bit can represent a quantity, 'zero' or 'one'.

If we have eight bistables, their eight outputs are each represented by '0' or '1'. The collection of eight bits is called a **byte**. One possible set of outputs is shown in the drawing in the next column. There, the byte is 10100110. This might represent eight different facts, such as 'It is raining', 'It is not windy', It is Monday', 'It is not a public holiday' and so on.

Alternatively, the byte might represent a quantity. '10100110' is the binary equivalent of the decimal number 166. This could be the output of an 8-stage counter and '10100110' (or decimal 166) is the number of cars in an automatic car park.

It is important to read the outputs in the right order.

In this example, we are reading from bottom to top. The bottom bistable outputs the most significant bit (MSB). The top bistable outputs the least significant bit (LSB).

A byte, when representing number, can take any value in the range 0000 0000 (all bistables reset) to 1111 1111 (all bistables set). This is equivalent to 0 to 255 in decimal, a total of 256 possible values altogether.

Multiples

The memories of most of today's systems need to store more than a few bytes of data. Larger amounts are rated in:

*   **Kilobyte** (Kb): A kilobyte is 1024 bytes. Note that it is *not* 1000 bytes, as one might expect from the meaning of 'kilo' in the metric system. A kilobyte of memory is enough to store a short document, such as half a page of text. It could also store the instructions for operating a simple circuit such as an automatic dishwasher.
*   **Megabyte** (Mb): A megabyte is 1024 kilobytes. It is *roughly* equal to a million bytes, or 8 million bits. A megabyte could store a small colour photograph, or the instructions for a simple computer game.

- **Gigabyte** (Gb): a gigabyte is 1024 megabytes, or approximately a thousand million bytes. The total storage capacity of a personal computer is usually a few gigabytes.

Memory

Many automatic systems such as dishwashers, microwave ovens, robots, and computers are based on logic and need memory for data storage. As suggested above, the data may be in the form of instructions to the machine, telling it how to perform its tasks, or how to respond to given inputs. Data of this kind is called a **program**. We shall say more about programs later. Data may also be numeric, as in an automatic weighing machine.

Both program data and numeric data are stored in binary form. The memory chip carries thousands or even millions of bistables, each storing a single bit of data, '0' or '1'. The bistables may each be similar to that shown opposite but a different kind of bistable is found in many types of memory chip.

A typical example of a memory chip is the 6116, which stores 2 kilobytes of data.

The IC is a large one, but not because the chip inside is particularly big. The main reason for its size is to provide space for the 24 pins that are needed for the power supply and for input and output signals.

Data is loaded into the chip or recovered from the chip a byte at a time. This means that the chip must have eight output terminals, one for each bit of the byte. When data is being loaded (or, as we more often say, **written**) into memory we have to be able to tell the chip which set of eight bistables to store it in. This is easily done because each byte has its own **address**. This being a 2 Kb memory, the addresses are numbered from 0, up to 1 less than 2 Kb.

2 Kb is a little more than 2000 bytes. It is 2 times 1024 bytes, or 2048 bytes. So the addresses run from 0 to 2047. These numbers need to be in binary form for the logic circuit of the chip to be able to decode them (see p. 130). The binary equivalent of 2047 is 111 1111 1111. This is an 11-bit number, so we need 11 input lines to specify the address of the byte in which the data is to be stored.

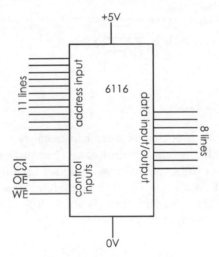

Two of the remaining five pins are used for the power supply to the chip. Finally there are three control inputs:

- **CHIP SELECT**: A system may have more than one memory chip. To make one particular chip operate, its $\overline{CS}$ line is made low.
- **OUTPUT ENABLE**: If this line is made low, the chip outputs data from an address in its memory.
- **WRITE ENABLE**: When this is low, data that is placed on the 8 data lines is stored at an address in the memory.

145

The bars over $\overline{CS}$, $\overline{OE}$ and $\overline{WE}$ mean that the lines are normally held at logic high, and have to be made low to act. We call this **active low**.

There are two kinds of operation, writing and reading. To write some data into memory, the CHIP SELECT input is made low. Then the byte to be written into memory is put on the data lines, and the address where it is to be written is put on the address lines. Finally, when the WRITE ENABLE line is made low, the data is stored in the eight bistables at the given address.

The diagram on the right shows logic levels at the moment when the byte 1001 1100 is being written into address 100 0110 0110. In decimal, the value 156 is being written into address 1126.

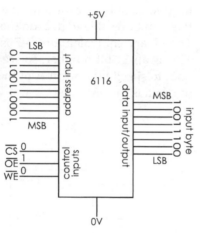

To **read** data that is stored in memory, make the CHIP SELECT input low. Then put on the address lines the address to be read from. The stored data appears on the data lines when the OUTPUT ENABLE line is made low. Below, we are reading from address 15.

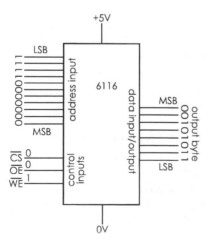

The 6116 and similar memory ICs are used for temporary storage of data. When the power supply to the circuit is switched off, the data is lost. Other types of memory IC are able to retain the stored data indefinitely. It remains there when the power is switched off. It changes only if new data is written into it. This is the kind of memory used in digital cameras, for example. The data for several dozen colour photographs can be stored in the memory card of the camera. Later, after the data has been transferred to a computer, or has been printed out on to paper, the data can be erased and a new set of photographs recorded.

PLDs

A **programmable logic device** is an integrated circuit that consists of a large number of logic gates of different kinds on a single chip. The connections between the gates can be programmed into the IC to build many different kinds of logic circuit. The PLD replaces a circuit board carrying different ICs with different kinds of gate and a complex network of connections between them.

A similar function can be obtained by using a memory chip as a PLD. For example, on p. 130 there is the circuit of a decoder that converts a 2-bit BCD number to produce the outputs needed to drive a 7-segment display. This decodes only the numbers 0 to 3. To decode all the numbers from 0 to 9 requires a much more complicated circuit with many more gates. We could build this circuit from logic gates or obtain the same function using a simple memory chip.

The memory chip would need only 10 addresses, one for each of the digits 0 to 9. At each address we could store the data needed to drive the display to produce the corresponding digit.

The MSB of each stored byte is 0, as we need only 7 bits to code the seven segments. After this, the data codes the segments from segment *a* to segment *g* (LSB). This data is permanently written to the chip.

Address (in decimal)	Address (in binary)	Data
0	0000	01111110
1	0001	00110000
2	0010	01101101
3	0011	01111001
4	0100	00110011
5	0101	01011011
6	0110	01011111
7	0111	01110000
8	1000	01111111
9	1001	01111011

To produce the required 7-segment code, we simply put the number of the required digit on the address lines, and make the OE input low. The diagram below shows how the digit '5' is produced (Note: we need only 4 address lines for addressing 10 bytes).

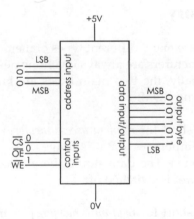

This is a simple example of how a memory chip can be used as a PLD. Generally, we would use it to perform much more complicated logic. For example it could drive a 5 × 7 LED display to show any numeric or alphabetic character.

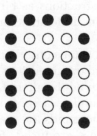

Questions on storing data

1 What is a bit and what values can it have?

2 How many bits are there in 2 bytes?

3 How many bytes are there in 5 Kb?

4 Describe a 1-bit memory circuit.

5 When a given bit is '1', it means 'The temperature is greater than 25°C'. What is meant when the bit is '0'?

6 What is the value of the most significant bit in the binary number 1 0011 0010?

7 Given the logic levels shown in the diagram, what operation is being performed by this memory IC?

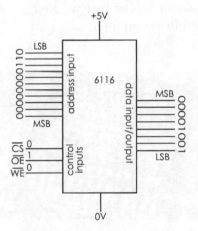

8 What are the three main groups of inputs or outputs of a typical memory IC?

9 What is the action of the CHIP SELECT input on a memory IC, and when is it needed?

10 What is the meaning of the bar over the names of input terminals such as WE and CS?

11 Describe how a memory IC can be used in place of a progammable logic device.

12 Give three examples of equipment or appliances that have memory built in to them. In each case, say what kind of data is stored in the memory. If possible, state the size of the memory.

13 Explain what is meant by the terms 'write' and 'read' when applied to the action of a memory IC.

33 Microcontrollers

Microcontrollers are not included in some Electronics specifications

All kinds of electronic equipment, from mobile phones to microwave ovens and from dishwashers to digital cameras, have a microcontroller at the heart of their system. The microcontroller performs all the complicated processing needed to link the inputs of the system to its outputs.

A microcontroller is sometimes referred to as a 'computer on a chip'. This is a good description. A microcontroller is a single integrated circuit on which are combined most of the circuit blocks that we find as separate units within a computer.

There are hundreds of different microcontrollers available. It is difficult to say that any one type is 'typical'. The 28-pin integrated circuit in the photo is average in size and contains the items that are needed by all controllers:

- **Arithmetic-logic unit (ALU)**: Logic circuits that perform addition, subtraction, and many logical operations.
- **Memory**: Logic circuits for storing data. There are two kinds. **RAM** stores up to 72 bytes of data. This memory is used by the ALU for a temporary store of data that it needs while processing. **ROM** stores up to 3 Kb of data. This is where the program is stored that tells the controller what to do.

- **Clock**: The chip contains all the components of the system clock (p. 135) except for the crystal.
- **Input and output**: Of the 28 pins of this IC, 20 are used for the input or output of data. They may be connected to sensors and other input devices. They may be connected to lamps, displays, motors, loudspeakers and other output devices. There is more on p. 150 about interfacing these to the microcontroller.

As well as these essential units, the controller in the photo has special facilities that many, but not all, microcontrollers possess. These include a pair of digital timers and an 8-stage binary counter.

Memory

There are many different types of memory, and manufacturers are always inventing new types. Essentially, the two main types are known as RAM and ROM.

RAM (short for *random access memory*) is used for temporary data storage, as already mentioned. The data stored there is lost when the power is switched off.

ROM (short for *read-only memory*) is more-or-less permanent. It is intended to be read from often but not often written to. It stores instructions as binary code that the controller can 'understand' and act on.

In some controllers, the data in the memory is programmed into it when the chip is made. It can not be altered. This type of ROM might be found in a controller for a washing machine. Thousands of ready-programmed ROMs are mass-produced for a particular model of machine.

Many controllers, such as the one on p. 148, have ROM in which the program can be erased and then re-written as many times as you want. The photo below shows a device used for programming the ROM of this kind of controller.

The programmer has circuits concerned with providing the voltages necessary for storing data in the ROM of the controller. Programming is controlled by a computer, on which the user creates and tests the program. The board has a special 20-pin socket into which a controller is plugged for programming.

When the program is complete and tested, the controller is removed from the socket. It is then put in a socket in the device (such as a home security system or an electronic kitchen scales) that it is intended to control. The 40-pin socket on the right of the controller is for programming the more complicated controllers that have more input and output pins.

Writing the binary code of the program into the ROM of the controller is under the control of software running on a computer. The person creating the program, types in instructions telling the controller which of its inputs to check on and what to do if the inputs are at logic high or low (depending on the sensors attached to the inputs). The program tells the controller what to do (that is what outputs to activate) when it detects certain combinations of inputs.

When programming software is run, the computer screen looks something like that at the bottom of the page.

The operator is typing the program into the window at top left. The operator uses a special 'language' to tell the controller what to do (more about that in Topic 34). The other windows tell the operator about the state of various parts of the controller. This includes a list of all the input and output pins and the present state of each of these. Using software such as this, the operator creates a program and tests it before actually writing it into the controller. The program can be run at its usual speed or just one step at a time. Stepping through gives the operator time to check that the program is working correctly at each step.

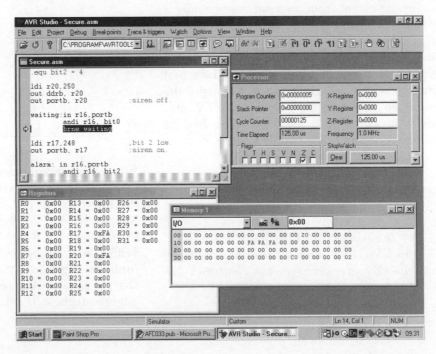

The operator can also decide on key stages of the program at which the run should stop. When the program stops at the breakpoint, the operator can study the contents of memory and registers, and the state of the outputs at the end of each stage.

Up to this stage, the controller itself has not been programmed. Everything is being done in the operator's computer. When the operator is satisfied that the program works correctly, it is downloaded into the ROM. The software in the computer converts the instructions keyed in by the operator into a binary code that the processor can 'understand'.

If the program is still not quite right, it is corrected on the computer and the new version downloaded into ROM, to replace the original version.

Input interfacing

An input pin of a typical controller can accept up to 25 mA. This does not mean that we can feed 25 mA into all 20 pins at once. That would be a total of 0.5 A, which would be too much for the chip to handle. The data sheets state maximum limits for input currents. Normally, the input current from a sensor may be only a few microamps, so the problem does not arise.

The other point about input to a microcontroller is that the voltage of the signal must be within a certain range. Normally the minimum voltage is 0 V, or close to 0 V, and this is accepted as a logic low. The highest voltage is normally the supply voltage, or close to it. This is accepted as a logic high.

Most microcontrollers run on a range of supply voltages. The PIC microcontroller illustrated on p. 148 runs on any voltage between 2 V and 6.25 V. The Atmel controller on p. 149 runs on 2.7 V to 6 V. These ranges allow the chip to be conveniently powered from batteries in portable equipment.

Inputs to the controller usually come from switches or sensors. Here are some examples:

Switches: This category includes pressure mats, microswitches, push-buttons and keyboard switches. Typical circuits are shown below.

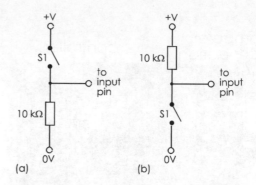

Circuit (a) normally gives a low input, but gives a high input when the switch is closed. Circuit (b) normally gives a high input, but gives low when closed.

Voltage dividers: Sensors such as thermistors and LDRs are usually a part of a voltage divider network. We use a version of the circuit on p. 80 to obtain a binary input .

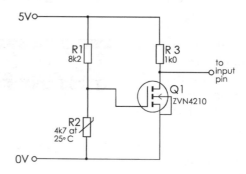

When the temperature is above a certain level, Q1 is off. There is no voltage drop across R3, so the input to the pin of the microcontroller is logic high.

When the temperature is below that level the input is logic low. Switching occurs fairly sharply as temperature crosses the set point. If a more precise action is required, use a Schmitt trigger, as on p. 103.

Another way of obtaining the same action is to use a microcontroller that has a comparator on the chip. The circuit is the same as that on p. 108, except that the comparator is included in the microcontroller. The output of the comparator sets or resets a bit in the memory, so the ALU can read '0' or '1' from this, depending on the temperature, and act accordingly.

Analogue input: For some projects we need to input the actual voltage across the sensor, instead of simply '0' or '1'. For example, we might need to record air temperature once an hour throughout the day, and also find minimum, maximum and average values. Although the input will actually be a voltage in the range 0 V to the supply voltage, there is no problem in converting this to a Celsius temperature. For analogue quantities we need an **analogue-to-digital converter**, or **ADC**. If you need this feature, check through the data sheets to find a controller that has one (or more) ADCs on board. The data sheet will tell you how to use it. Alternatively, use an external ADC such as the ADC0804LCN. This has an 8-bit output that can be fed directly to a microcontroller.

Output interfacing

The output pins of a controller can supply only a limited current. Typically, this can not be more than 20 mA. Outputs can burn out in a second or two if this limit is exceeded. As with inputs, there is an overall limit to the total output current that can be supplied at any one time.

Light emitting diodes: These can be driven directly, provided that there is a current-limiting resistor in series with the LED. To be on the safe side, limit the current to 10 mA. Given the supply voltage is 6 V, a 390 Ω resistor is suitable.

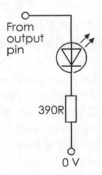

LEDs are useful for displaying output logic levels when testing a new project.

Transistor switches: Most output devices, such as motors, solenoids and loudspeakers, take such large currents that they are best driven by a transistor switch. A MOSFET switch is ideal (p. 80 and p. 82) as it takes virtually no current. It can not overload the output circuit of the controller.

The circuit on p. 82 shows how to interface a relay using a MOSFET switch. This is useful for controlling heavy-current devices. Smaller loads can be switched by reed relays, directly from the output pins.

Pulsed switching: On pp. 119-20, a method is described of varying the duty cycle of a 555 astable. This is used for lamp dimming and for controlling the speed of a motor. The same effect can be produced by a controller that has a built-in pulse generator. Pulse width is set by programming the controller.

Questions on microcontrollers

1 Name three appliances or pieces of equipment that contain a microcontroller.

2 What are the main units present in a microcontroller, and what does each do?

3 What are RAM and ROM? Describe the main uses of each of these.

4 Using a data sheet, list the amounts of RAM and ROM present in a named microcontroller.

5 In what type of applicance or equipment would you expect to find permanently programmed ROM.

6 Outline the process of programming a controller from a computer.

7 Describe how you would interface **(a)** an LDR and **(b)** a tilt switch to a controller.

8 Why is it essential to know the current needed to drive an output device?

9 Describe how you would interface **(a)** a 6 V 360 mW filament lamp and **(b)** a mains-voltage electric fan to a controller.

34 Programs

Programming is not included in some Electronics specifications

The **program** stored in the ROM of a microcontroller tells it what to do. It consists of a series of **instructions**. An instruction might *mean* 'Find out the logic level at input pin 5'. But, although this is what the controller will do when it comes to that instruction in its ROM, the instruction is not there as a sentence made up of words. It is coded, often as a single byte of a binary code.

To help you understand what a program really is, we will look more closely at a very short section of an actual program. It runs on the controller pictured on p. 149. It is part of the security program featured in the screen shot on the same page. Follow the descriptions, noting the main features that are printed in **bold type**.

If we were to check the contents of the ROM in which the program is stored, we would find this:

Address	Byte
0	11101111
1	01001010
2	10111011
3	01000111
4	10111011
5	01001000
and so on ...	

This is what the controller has to read, byte by byte. The '0's and '1's represent bistables that are either reset or set. This is what the programmer has to put into the ROM in order to make the controller do anything useful. This array of '0's and '1's is called **machine code**. Few operators try to program directly in machine code.

Summing up:

A program contains a series of instructions, coded by the states of bistables in ROM.

Codes for programming

The screen shot on p. 149 shows the program being written by using a special kind of software, called an **assembler**. The program begins with the line:

ldi r20, 250

This too is a code but a little easier to understand than machine code. The first part of the code 'ldi' means 'Take the value that comes next and *load immediately* into a specified register'. The byte stored at address 0 is the machine code version of this instruction. This means something to the controller but not to us!

The second part of the code 'r20, 250' means 'The value is 250 and the place to load it is register 20'. This data is coded in the byte at address 1. Register 20 is one of 32 sets of 8 bistables that form part of the RAM. So the first two bytes of the program tell the controller to put the value 250 in register 20. Register 20 will then look like this:

11111010

This binary number is the equivalent of decimal 250. Note that byte 1 is not an instruction. It is coded data: what to put and where to put it.

As well as instructions, a program contains coded data.

The next part of the program sets up the controller to operate the security system. We will not go into the details of how bytes 2 to 6 code this. After that, we have the instructions for waiting for input and producing output.

A program written in assembler lists an instruction, and often a byte of data, for every operation that the controller does. There is no way of combining the two into one statement. The instructions all specify very simple, short processes. For example, when two numbers are to be be added together, there must be instructions to place the numbers in different registers, then add them, then transfer the result to some other register.

Processing is broken down into very elementary steps.

As a result of this, a program has many instructions, and takes a long time to write.

Other codes, or languages, are often used for programming, using different software. Some use BASIC, which is easy to learn. The instructions are written in statements that are very like ordinary English sentences. For example, the instruction on the opposite page would be:

LET r20 = 250

This is equivalent so several lines of machine code or assembler code. Also it is much easier to understand.

Another language that is often used for programming controllers is called C. This has many useful features though it takes longer to learn. Summing up:

Programming is done in assembler or a high-level language, such as BASIC or C.

Inputs and outputs

The controller shown on p. 149 has a set of 8 pins for input or output. These can be programmed single or as a byte. A set of pins such as this is often called a **port.**

In this very simple programming example, we need only three bits. Two are programmed to be inputs (from the sensor and from the reset button). One is programmed as an output (to switch on the siren).

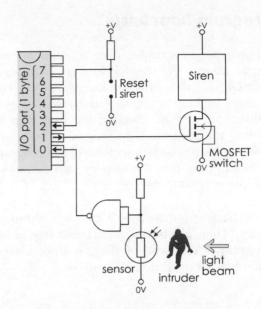

The sensor circuit includes a NAND gate with its inputs connected together. It acts as a NOT gate. The output goes low when the beam is broken by the intruder, making bit 0 low.

The output bit 1 is normally low. When it is made high it switches on the MOSFET and the siren sounds. Bit 2 is an input that reads the state of the reset button. Pressing the button makes bit 2 go low.

There are 5 unused bits on the port and the controller also has another port with 7 bits. We could use these to control a more complicated security system. In fact, it is barely worth while to use a controller on such a simple system as our example. The same action can be obtained with a few logic gates. However, the point of this example is to show that a system built up from **hardware** (logic gates, op amps and other electronic components) can be mainly replaced by **software** (programs run by a controller).

The advantage of software over hardware is more easily seen when the system is a complicated one. Designing, writing and possibly amending software is much easier than wiring hardware, especially if the hardware needs to be re-wired to correct it or improve it.

Program flowcharts

To program a controller, you will need to learn more about the assembler or high-level language of the programming software. The manual or help screens of the software will guide you in this. Usually, the first step in writing the program is to prepare a flowchart. This outlines the main sections of the program without going into the step-by step details of the instructions to the controller.

Flowcharts are drawn as a number of linked boxes. There is a box for each major step of the program. The shape of the box depends on what kind of operation is involved:

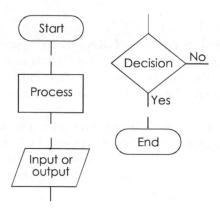

To illustrate the way these are used, on the right is the flowchart of the security system referred to on p. 152.

The program begins by setting up the three pins as two inputs and an output. This is a *process* within the controller. Next, the controller inputs the state of bit 1, the output from the sensor. If it is not '0', the light beam is unbroken. This is a decision box, and the answer is NO. The program loops back to read the input again. It does this repeatedly, looping round and round, waiting for an intruder to break the beam. This is a **wait loop**.

If the beam is broken, the answer to the question in the decision box is YES. Then the controller continues to the next step of the program. Output at bit 1 is made high, turning on the siren.

The siren continues to sound while the controller waits in the next loop. This time it is inputting data at bit 2, the state of the reset button. It waits until the button is pressed and bit 2 becomes '1'. Then the answer to the lower decision box is YES. The controller loops back almost to the beginning of the program. There it turns off the siren and waits in the first wait loop for the next intruder to break the beam .

Note that this program runs continuously in loops and does not have an END box.

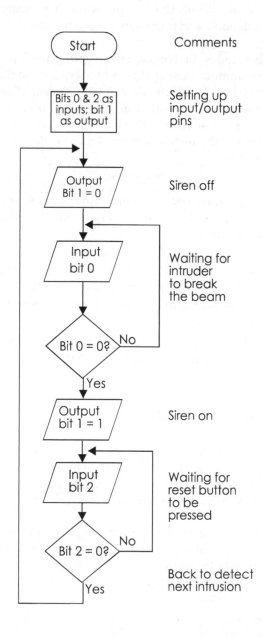

Automatic doors

The flowchart on the right is the first step at writing the program. Note that the flowchart shows the software. The hardware is shown in the system diagram on p. 88.

The flowchart begins with the usual setting-up of input and output pins. Then the doors are closed, if they are not already closed. It is usually safer to begin a program with an action such as this, to ensure that the system is in a specified state to start with.

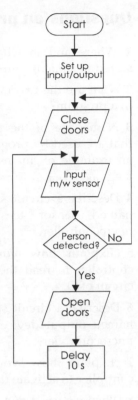

Next, the input from the microwave sensor is read. If a person is detected approaching the doors, the program runs on to open the doors. If not, the program loops around, reading the sensor repeatedly until someone approaches. After the doors are opened, there is a delay to give the person a chance to pass through. After that, the program jumps back to read the sensor again. The person may have been walking too slowly or a second person may be approaching. In these cases the doors remain open. However, if no one is approaching, the doors are closed.

It is a good approach to a program to map out the main stages first. Details follow later. In this example, it is important to program the opening and closing of the doors more carefully. A practical point is that the doors do not need opening if they are already open. Another point is that just switching on a motor for a given time does not guarantee that the doors are fully open. We need positive input about whether the doors are fully open or not. This data about what has (or has not) happened is known as **feedback**.

The system needs two more inputs, perhaps from two microswitches. One closes when the door is fully open and the other closes when the door is fully shut. Using these sensors, we expand the 'Open doors' stage of the previous program.

The first step is to check if the doors are already open. If so, the program skips to the end of the open doors routine. The motor is turned on to open the doors. Then the sensor is read to find out if the doors have opened as wide as possible. If not, the routine loops around, with the doors motor still running, to check again. When the 'doors-open' sensor confirms that the doors are open, the motor is switched off. This is an example of feedback.

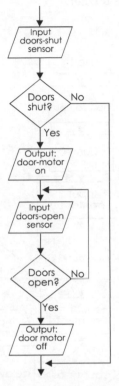

A matching routine should be added to the main program to supervise the closing of the doors.

Things to do

Even this is not the end of the programming. What happens if someone leaves a supermarket trolley in the doorway as the doors close? How can we provide a manual over-ride button to open or close the doors at the beginning and end of the day, or when the window cleaner needs to work on the doors? It would also be a useful feature to have automatic locking, so that the doors can be made secure when the supermarket is closed. There are lots of refinements possible. Can you think of more? Write the programs.

Twisty wire game

The microcontroller version of this game (p. 121) has one input. This goes high (=1) when the loop contacts the twisty wire. It has one output, to a MOSFET switch that turns on a buzzer.

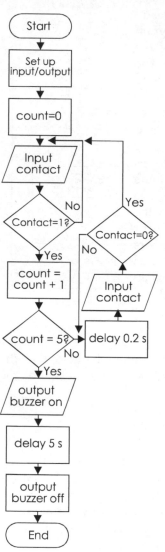

The program begins by setting up the input and output pins. Then a register, called *count*, has zero loaded into it. *Count* has 1 added to it every time contact is made between the loop and the wire.

The player is allowed five 'contacts'. When *count* reaches five, the buzzer is switched on for 5 seconds and the game ends.

Note the 0.2 s delay. This allows for the **contact bounce.** Multiple makes and breaks of contact are registered as only one contact on each occasion. It gives time for the bounces to settle.

Contact debouncing can be done by using a capacitor (hardware debouncing) but here we use **software debouncing.** The routine used here also checks the input until it has returned to '0' before returning to wait for the next contact.

Questions on programs

1 Where and in what form is the program stored in a microcontroller?

2 What are the two main kinds of information in a program?

3 Name one of the programming languages that are used by programming software. Give an example of an instruction written in that language.

4 Describe a circuit suitable for interfacing a named sensor to an input pin of a microcontroller.

5 Explain how you would instruct the controller to read the input from the sensor circuit of Q. 4.

6 Describe a circuit suitable for interfacing a named output device to an output pin of a microcontroller.

7 Explain how you would instruct the controller to activate the output device.

8 When we use a microcontroller, much of the hardware of a circuit is replaced by software. What are the advantages of this?

9 Describe the meanings of the shapes of the two flowchart boxes drawn below:

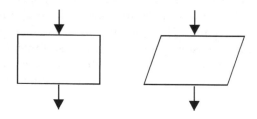

10 Why is a flowchart the best first step toward writing a program?

11 What is the difference between a program flowchart and a system diagram?

12 What are normally the first processes at the start of a flowchart?

13 Draw a flowchart for a panic button program, operating as described on p. 121.

14 Draw a flowchart for the operation of a washing machine (with only one washing programme).

Design time

Design programs for the systems described on this page. Draw the flowchart and, if your course includes this, write and debug the program. If possible, program a microcontroller and attach the hardware needed to complete a fully operational system.

Make the twisty wire game more competitive. Extend the program to put a time limit on the player's turn.

The twin-loop model railway system on p. 133 can be made more realistic by putting it under the control of a microcontroller. It could route the train in several different ways and perhaps use a more complicated layout. You may want to add extra sensors. If you have the model equipment, you could add electrically operated signals. This makes an effective Open Day exhibit.

Make the twisty wire game more rewarding. Add a special contact sensor at the far end of the wire. Then extend the program to flash LEDs when the player makes a clear run.

Add to the realism of the model railway by designing a circuit sensitive to the sound of a whistle. Program the controller to start the train when the whistle is blown once and to stop it when it is blown twice.

Microcontrollers are ideal for programming lighting displays. Connect MOSFET switches (or possibly relays) to half-a-dozen or more output pins. Then wire up panels or strings of lamps. Program the controller to produce a scintillating display. Fun at Christmas time.

Devise a reaction tester program for the system described on p. 143. Then extend it to include other features that you would like it to have.

The intruder system on p. 123 can be implemented and extended by programming a microcontroller to drive it. Most controllers can perfom the AND operation so the hardware AND gate can be replaced by its software equivalent.

Design and program a weather station system. This could be done by several students, each working on a different section. It could show present temperature, and 24-hour maximum and minimum. You might design simple sensors for wind direction, wind strength, pressure and rainfall.

Design a metronome system based on a microcontroller. Devise a system of lamps or sounds ('clicks') to emphasise the beat at the beginning of each bar. Try programming Latin American rhythms.

Build and program a working model lift or car-park entrance gate.

35 Visual output

Apart from the video tubes used in oscilloscopes, TV sets and radar displays, the most important visual outputs are based on lamps, LEDs, and liquid crystal displays.

Lamps and LEDs

LEDs are probably used more widely than any other device for indicating the output of a system. They are very efficient, use little current, last much longer than filament lamps, are less easily broken. They are made in a wide range of colours, shapes and sizes. They are also available with a built-in flashing circuit. Because they take little current, they can be driven directly from the output of a 7555 timer IC, TTL, CMOS buffer gates (4049 and 4050), microcontrollers and many other devices. This simplifies circuit design and construction. Their main disadvantage is that they are not easy to see when viewed in bright sunlight. However, really bright LEDs are being developed, so this will not be a problem in future.

Filament lamps are used where brightness is important. But brighter light means higher currents and filament lamps nearly always need a transistor switch to drive them.

7-segment displays

These are described on pp. 130 and 140. The 4511 CMOs decoder described there has outputs capable of driving the segments of an LED display. The outputs are active-high. This means that when a segment is to be switched on the corresponding output goes high. The display used with this IC must be the common cathode type (p. 140).

A decoder-driver from the TTL family is the 74LS47 IC. This has four BCD inputs and seven outputs to the segments. The outputs are active-low and need a common-anode display.

The anodes are connected inside the display and to a terminal pin that is connected to the positive supply line. The individual cathodes are connected, through series resistors to the outputs of the 74LS47.

Liquid crystal displays

An **LCD** consists of a special fluid sandwiched between two sheets of glass. There is a **backplane**, which is a transparent electrode coating the inner surface of the rear glass. Normally the liquid is clear. Regions of it appear black when an alternating voltage is applied between the backplane and the pattern of transparent electrodes that is plated on to the inner surfaces of the front glass. LCDs are usually made with several digits and decimal points. Often there are extra symbols and words to suit the LCD for special applications. The digital multimeter shown on p. 14 has an LCD. There are hundreds of other applications for LCDs, including pocket calculators, kitchen scales, microwave ovens and CD players.

LCDs can be driven by the BCD-to-7-segment ICs already described. However, it is simpler to use the 4543 LCD driver. Like the other ICs, it includes a BCD decoder with 7 outputs. There is an input for the **display frequency**. This is a square wave at about 200 Hz, generated by an astable. When a segment is to appear black, the IC generates an alternating field between the appropriate electrodes.

LCDs have the advantage that they are easy to read in bright daylight. In dull light or darkness, there must be a lamp mounted behind the display. A big advantage of LCDs for portable equipment is that the LCD takes only microamps of current (compared with about 20 mA *per segment* for an LED display). Clock, watches and thermometers with LCDs run for months on a small button cell.

36 Audible output

Alerts

A piezo-electric **buzzer** produces a low-pitched sound suitable for a door alert.

The buzzer on the left operates on 6 V and requires 20 mA. Its power leads are coloured red and black. Connect red to positive and black to negative. There are two fixing lugs. Bolt the buzzer firmly to the case or circuit board to obtain maximum sound.

A small piezo-electric **siren** produces a penetrating high-pitched note. The one below operates on 3-16 V and requires only about 5-7 mA.

The note is continuous but is made much more noticeable by sounding it in short bursts. Drive the siren from an astable running at about 1 Hz.

The sirens can also be used as alerts or alarms. Their sound level is between 100 dB and 110 dB. For maximum loudness, the siren must be firmly attached to the enclosure or circuit board.

A **sounder** is a brass disc coated with piezo-electric material. It may be driven directly by a 3 V peak-to-peak square wave generated by CMOS logic. It is useful for producing a variety of alert sounds. Sounders are made in different diameters with different resonant frequencies. For maximum effect, the sounder should be mounted firmly on the enclosure or circuit board and be driven by a signal close to its resonant frequency.

Alarms

These produce much louder sounds than alerts. Also the sound itself is usually an intermittent bleeping or warble to catch attention.

A wide range of alarm sirens is available. The multiple 'screecher' (above) produces a warbling tone. It requires 50 mA at 6 V or (for a louder sound) 200 mA at 15 V. This siren is for indoor use but weatherproof sirens are made for use outdoors. Alarm sirens produce about 130 dB of sound. Very loud sirens such as these can damage your hearing if you are close to them.. Because of their loudness, they are effective at forcing an intruder to leave the premises as quickly as possible.

Loudspeakers

These can be used as alerts and alarms. An example of such use is a 'musical chimes' door alert. The photo shows a suitable miniature speaker, 26 mm in diameter.

The mylar cone of this speaker makes it weatherproof. The coil resistance of speakers is typically 8 Ω, so a power transistor is needed in the output circuit. For maximum sound, mount the speaker on a baffle, that is, a panel with a circular aperture cut in it to accept the speaker.

37 Mechanical output

Motors

Small DC motors are used for driving many mechanical projects. They normally run on low voltages, such as 6 V or even as little as 1.5 V. They usually require a few hundred milliamps to drive them. This means that they must be switched with either a power transistor switch or a relay.

Page 65 shows a diagram of a r e v e r s i n g switch, based on a DPDT switch.

The DPDT switch can be replaced by a relay that has two changeover contacts (p. 82). This allows the flow of current through the motor to be reversed. A second relay with single normally-open contacts is used to switch the motor on and off.

Small motors usually run at several thousands of revolutions per minute. Gearing can be used to reduce the rate of revolution and to increase the turning force. As an example, the motor in the photo has a worm gear on its shaft. This can engage with a large cog-wheel to act as a reduction gear.

Another way to obtain reduced speed and also to control the speed is to drive the motor with pulses of variable duty cycle. Extension box 59 on p. 120 shows the circuit.

Solenoids

A solenoid consists of a **coil** in which slides a soft-iron **plunger** (photo, top right). The plunger normally has just one end of it in the coil. When a current is passed through the coil, the plunger is pulled strongly into the coil.

The arrow shows the direction of forceful motion. The outer end of the plunger may be coupled to parts of a mechanism.

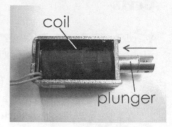

Note that this solenoid can *pull* but it can not *push*. Most mechanisms need to be able to return to their initial position. This can be done by using a spring or a rubber band. In some cases we can rely on gravity to pull the plunger back out of the coil. A third solution is to have two solenoids, one to pull the mechanism one way and the other to pull it back again.

In this photo, the plunger is on the left. There is a spring to force the plunger back out of the coil when the current is turned off.

Like the previous solenoid, this solenoid can be used to pull. However, there is a rod attached to the plunger. This projects on the right. When the current is switched on, the plunger is drawn into the coil and the rod *pushes* forcibly toward the right.

Solenoids require several hundred milliamps to power them so are best controlled by power transistor switches or by relays.

Other devices

These are often driven by solenoids. They include valves, taps, and door latches. A stepper motor is a useful way of obtaining controlled rotary motion.

Questions on output devices

1 List three visual output devices and suggest one use for each.

2 What is a seven-segment display? Mention one type of IC that can be used to drive such a display from a BCD input.

3 What are the advantages and disdvantages of an LCD compared with an LED display.

4 What audible output device is most suitable for use with:

(a) a cooking timer?

(b) a 'musical chimes' doorbell?

(c) a smoke alarm?

(d) a circuit that detects when a customer has entered a shop?

5 Describe the action of a solenoid. What type of circuit is needed to interface it to CMOS logic?

6 Design a relay circuit to start and stop a small DC motor and change its direction.

Design time

Design and build a minute timer. Use a 7555 astable to time the minutes and a CMOS counter to count the minutes from 0 to 9. Add a decoder and single-digit LED display to show the time elapsed.

An automatic plant-pot watering circuit turns on a water tap for 1 minute twice in every 24 hours. Design and build the circuit, including the watering mechanism. Use a ready-made valve or devise a water tap actuated by a solenoid.

Design and build a two-tone alert circuit using a 7555 astable and a frequency divider based on CMOS ICs. The 7555 would run at about 2 kHz. The divider produces a range of lower frequencies. Use two of these frequencies and design CMOS gating circuits to send these to a suitable audible output device.

Devise a solenoid actuated mechanism to open and close a greenhouse window. Add a circuit to open the window when the temperature inside the greenhouse exceeds, say, 25°C. The circuit is to close the window when the temperature falls below, say, 20°C.

A garage lamp is to be switched on when a sensor detects the headlights of an approaching car at night. There is to be no response during the day.
The light is to be switched off 5 minutes later, unless a push-button is pressed. A warning bleep is to sound 20 seconds before the lamp goes out. Another bleep is heard if the button is pressed. This circuit might need a microcontroller.

Design and build a one-way intercom system, with an alert to call someone to the receiving station.

Search the data sheets to find out about stepper motors and how to control them with a microcontroller. Design a project that is based on a stepper motor.

Design a robot.

38 Audio systems

The term 'audio system' could describe any system operating at audio frequencies, from a basic intercom to the sound reproduction system of a super-cinema. In this book, we look only at a typical stereo system for use in the home.

A system diagram (right) shows the main features of such a system. Typical inputs to the system are on the left.

Radio tuner: This receives radio signals carrying audio signals and converts them into electrical signals. The way it does this is described in more detail in Topic 40.

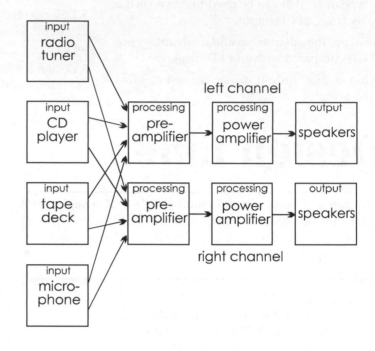

CD player: Music and sound, recorded digitally on a compact disc, are converted to analogue electrical signals by this unit. The disc consists of two plastic layers sandwiched together with a silvered coating between them (left).

The data is recorded as a series of dimples with spaces between them. The dimples represent logic '1' and the spaces between them represent logic '0'. As the disc spins, a beam of light from a low-power laser is focused on its underside. If there is no dimple, the beam is reflected back and detected by a photodiode. Where there is a dimple, the beam is scattered sideways and is not detected. This provides a stream of bits, either '0' or '1'. This is processed by complex logic circuits, eventually producing two analogue signals for the left and right stereo channels.

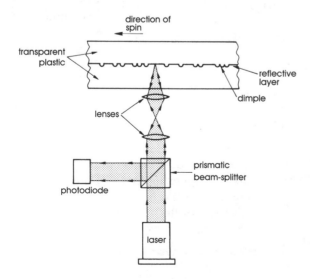

Tape deck: Audio tape is coated with a layer containing a magnetic substance such as chromium dioxide. This becomes organised into microscopic regions known as **domains**. Each domain is equivalent to a very small magnet. In an unrecorded tape, the domains are arranged irregularly, so there is no overall magnetization.

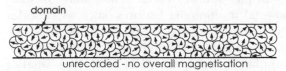

domain

unrecorded - no overall magnetisation

When sound is being recorded on a tape deck the tape passes a gap in a magnet in the recording head. A signal from an amplifier causes an alternating magnetic field in the gap, and this causes the domains to change direction.

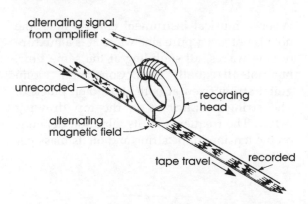

alternating signal from amplifier

unrecorded

recording head

alternating magnetic field

tape travel   recorded

In some regions they are mainly pointing one way ($\rightarrow$ below). In other regions they are mainly pointing the other way ($\leftarrow$). The directions and the proportions of domains affected correspond to the waveform that is being recorded. The diagram below shows the original analogue audio signal and the corresponding arrangement of the domains.

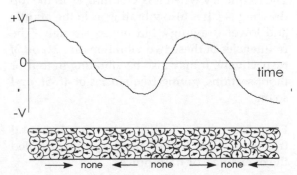

+V

0

time

-V

none $\leftarrow$ none $\rightarrow$ none $\leftarrow$

When the tape is played back, it passes under the playback head. There the magnetic fields produced by the domains on the tape induce alternating currents in the coil. These currents are a reproduction of the original signal current.

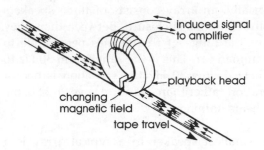

induced signal to amplifier

playback head

changing magnetic field

tape travel

Microphone: These are described on p. 97. Two microphones are needed to produce the left and right stereo signals.

Processing

There are two identical processing channels for the left and right stereo signals. The **preamplifiers**, accept the signals from the input stage that is switched through to them. These are **voltage amplifiers**, as described in Topic 28. The amplitude of signals from the input stages is usually no more than a few milliamps. Typical supply voltages for amplifier circuits is 15 to 30 V, and amplification increases the signal amplitude in this range. Currents in these amplifiers are small, to avoid generating noise during the amplification of the low-voltage input (p. 109). It is essential to avoid introducing noise at the early stage, when the signal is small, for the later stages will amplify the noise along with the signal. There may be several stages of voltage amplification.

The final stage of amplification is performed by **current amplifiers**. Current amplifiers often have no voltage gain, but the current gain may be 100 or more. The combined effect of amplifying voltage and then current is a high amplification of power. ($P = IV$, p. 48). The loudness of the sound produced is proportional to the power.

Output

The output from the power amplifiers goes to two sets of speakers. A single speaker for each channel is all that is essential, but quality systems have twin arrays of speakers. Typically, an array consists of three speakers mounted in one enclosure. Between them they cover the range of frequencies perceptible to the human ear. This is the range from 30 Hz to 20 kHz. Below 30 Hz the sensation is one of vibration rather than sound. Above 20 kHz the ear hears nothing.

The smallest speaker in a typical array is a tweeter, for sounds in the highest frequency range, 2 kHz to 20 kHz. The mid-range speaker has a greater diameter and covers the range 50 Hz to 5 kHz. The bass speaker, or woofer, deals with the low frequency range 30 Hz to 800 Hz. There may also be a fourth speaker, a sub-woofer, that handles 20 Hz to 200 Hz.

The signal from the power output amplifier is fed to these speakers through a cross-over network. This feeds each speaker with signals in its frequency range.

Other connections

A home audio system is most often used for listening to sound from one of the four input units shown in the diagram on p. 162. However, other connections may sometimes be used. For example, the signal from the radio tuner may be sent direct to the tape deck. This allows a radio programme to be recorded on tape to be listened to later.

Other inputs devices may also be attached to the system. For example, an MP3 player can be used as a source of recorded music. The term MP3 refers to a technique for compressing digital music files so that they become small enough to be stored in a reasonably small memory chip. MP3 files may be downloaded from sites on the World Wide Web, often for no charge.

The development of cheap 'flash' memory chips has made MP3 players affordable. This has made MP3 and more recent systems very popular .

Digital versatile disks (DVDs) are another possible audio source. Although they are more frequently used for recording films with multiple sound tracks. They may include special multimedia features, such as subtitles in several langauges. DVDs can be used for audio recordings of high quality. They are very similar to compact discs but have smaller dimples, more densely packed, and so store a much larger amount of data (up to 17 Gb compared with 650 Mb on a compact disc).

Musical sounds

When a musical instrument emits a note, the note is not just a pure sine wave. It is a mixture of sine waves, all sounding at the same time, but not all equally loud. Consider a violin guitar string fixed at both ends, then plucked. The string vibrates as in the top drawing below. The frequency of its vibration depends on the tension in the string and on its mass per metre.

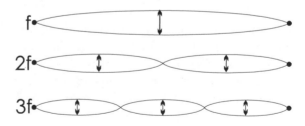

The frequency when it is vibrating as in the top drawing is f. It is also vibrating as in the middle and lower drawings, but not as strongly. The frequencies of these two vibrations are $2f$ and $3f$ respectively. It may also be vibrating in four or more sections, giving frequencies of $4f$, $5f$, and so on.

Frequency f is called the **fundamental** frequency. The higher frequencies are called the **overtones**, or **harmonics**.

If we use a microphone to pick up the sound made by the vibrating string and display it on an oscilloscope, we see the waveform labelled '*f* + 2*f* + 3*f*' in the drawing below.

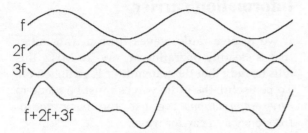

This is the sum of the fundamental note of the string plus the first two harmonics, drawn separately. The fundamental and harmonics are all sine waves of different amplitudes. But the frequencies of the harmonics are multiples of the fundamental. If we alter the length of the string to get another note, we get a different frequency and a different but related set of harmonics. In all cases, the combination of fundamental plus harmonics has the characteristic sound of a guitar.

Similarly, the air in a flute vibrates in several different ways, producing a fundamental and a different set of harmonics. Their relative amplitudes may differ and some may be missing. They sound like a flute when added together. Differences between the sets of harmonics are not the only features that make a flute sound different fron a guitar but they are an important difference.

The **bandwidth** (p. 107) of an audio system has a big influence on the quality of the sound it produces. The amplifier that gave the graph on p. 107 has a bandwidth from 800 Hz to 5 MHz. This might be suitable as a radio-frequency amplifier, but it is not suitable for audio frequencies. It does not amplify the lower frequencies. The graph shows that the signal is much reduced in amplitude between 30 Hz and 800 Hz. The bass response of the system would be very faint indeed. Ideally, the bandwidth of an audio system should extend from 30 Hz (or a little lower) up to 20 kHz.

Within the bandwidth, the amplifier should amplify all signals equally. In other words, the top of the curve should be flat, as in the diagram on p. 107. Then the system will reproduce the original sound with the loudness of all its harmonics in the correct proportions. Even though a musical instrument does not play fundamental notes as high as 10 kHz, some of its harmonics may be over 10 kHz. If these are not reproduced at their correct relative volumes, the 'mix' of harmonics in the sound is wrong, and the quality of reproduction is poor.

Tone control

Ideally, the outputs from the speakers is an exact replica of the original sound. The amplifier and all other stages in the system must have a bandwidth extending over the complete audio range. Unfortunately, there may be stages in the system in which the bandwidth is too narrow, or the frequency response is not flat-topped. For instance, the recording microphone may emphasise some frequencies and reduce others. The amplifier, the speakers and their enclosures, and even the furnishings of the room exert their effects on the final sound as heard by the listener.

To compensate for these effects, most audio systems have **tone control** circuits (above). These consist of capacitors, resistors and op amps. They allow bands of frequencies to be boosted or cut. The frequency response can thus be adjusted to give high-fidelity sound reproduction.

39 Radio transmission

If we apply the output from an oscillator to a pair of metal rods, electrons rush to and fro along the rods. Their rapid motion generates an **electromagnetic field**. Electromagnetic waves spread outward from the rods, like ripples on a pond, only in three dimensions. If the frequency of the oscillator is between 30 Hz and 30 GHz, the waves are **radio waves**.

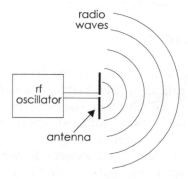

The paragraph above describes a basic **radio transmitter**. In the case of the transmitter we have described, its **antenna** or **aerial** is the pair of rods (or **dipole**), but it may be a piece of wire several metres long, suspended from masts or a similar structure. It may have a reflector of metal rods or wire mesh to concentrate the radio waves into a beam.

Radio waves travel from the transmitting antenna. They may travel for millions of miles through Space, as when we communicate with a space probe approaching the planet Pluto. However, not all transmissions travel so far. Depending on their frequency, radio waves may be reflected back toward the Earth's surface from layers high in the atmosphere. Others stay fairly close to the Earth's surface.

Focused beams of radio waves are usually directed at a receiver that is a few tens of kilometres away. The receiver is in line of sight of the transmitter. Radio waves that are reflected in the earth's upper atmosphere may be detected at distances of many hundreds of kilometres fron the transmitter.

Information carrier

If we receive radio waves from a transmitter that is steadily operating at, say, 10 MHz, this tells us only that the transmitter is switched on. To be useful, the radio waves must be made to *carry information*. We use the basic radio-frequency as a **carrier wave**.

One way to carry information is to switch the output stage of the transmitter on and off. The transmission is a series of short bursts or **pulses** of the carrier.

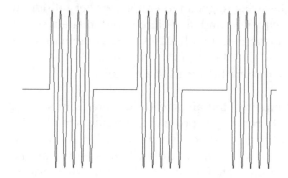

We say that the carrier is **modulated** into pulses. The pulses are all the same amplitude but the pulses and the gaps between them may vary in length. The pulses are a code, which represents the information that is being carried. This type of transmission is known as **pulse code modulation** (PCM). It is the most widely used technique for carrying digital information, and there is more about this in Topic 41.

When a PCM transmission is received, it is first **demodulated**. The radio-frequency waves are removed, leaving the pulses. These are sent to a decoder, to convert them to understandable data. As well as being widely used for data transmission, PCM is used in remote control systems, such as car immobilisers and domestic security systems.

Amplitude modulation

AM is used for transmission of analogue signals. Before transmission, the carrier wave is passed through a modulator. This modulates the *amplitude* of the analogue signal on to the carrier wave.

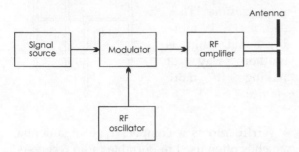

The result is shown below:

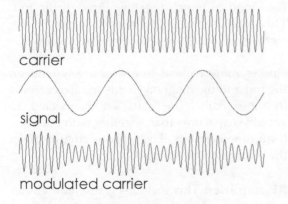

The modulated carrier is transmitted. It is demodulated at the receiver (p. 168).

Frequency modulation

FM is an alternative technique for placing an analogue signal on a carrier wave. We modulate the *frequency* of the carrier according to the analogue signal. The amplitude of the carrier is constant. The effect of FM on the carrier is shown below:

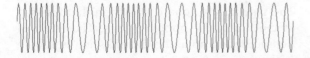

The frequency increases as the signal voltage goes more positive. It decreases as it goes more negative. To make the effect clear, the diagram shows a large increase and decrease in frequency. In practice, the variation in frequency is much less than this.

FM is not affected by changes in the amplitude of the modulated signal. This means that if reception is poor and the amplitude varies, the strength of the demodulated signal remains unaltered. Another related advantage is that sudden spikes in the amplitude of the signal, caused perhaps by lightning discharges or local electrical disturbances, have no effect on the frequency. There are no crackling sounds of interference such as are common with AM transmission.

Questions on audio systems and radio transmission.

1 Draw a system diagram of a domestic audio stereo system. Explain the function of three of its parts.

2 What is the principle of a CD player?

3 Outline the way in which a tape recorder and player work.

4 Why does the pre-amplifier amplify voltage and not current?

5 What are the lowest and highest audio frequencies?

6 Explain the meanings of the terms 'fundamental' and 'harmonic'.

7 What is meant by the term 'bandwidth'? With the help of a diagram describe the frequency response of an ideal audio amplifier.

8 Why is tone control needed in an audio system?

9 Explain how an oscillator is made to produce radio waves.

10 What is meant by the terms 'carrier wave' and 'PCM'.

11 Describe one way of transmitting analogue information by radio.

40 Radio reception

Sensitivity and selectivity

These are two important features of a radio receiver.

A *sensitive* radio receiver is able to pick up signals from weak or distant transmitters.

A *selective* radio receiver is able to pick out the signal of a particular transmitter from the signals on close-by frequencies that are arriving from other transmitters.

Radio receiver

The diagram shows the main stages in a tuned radio frequency receiver.

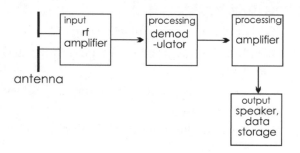

Antenna: This may be a **dipole**, similar to the antenna of the transmitter. Other types of antenna may be used, for example, a **long wire** suspended well above the ground on two or more masts. When radio waves strike the antenna, their electromagnetic field makes the electrons in the antenna oscillate.

The electrons in a dipole antenna oscillate most strongly when receiving waves of a particular frequency. This makes the antenna more sensitive to transmissions of a given frequency. It is more *selective*. It helps to tune the system to receive a particular station.

Dipoles often have extra elements to make an array that is more *sensitive*. They also make it more directional, which is another way of making it more *selective*.

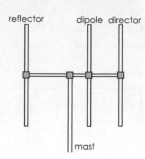

A **ferrite rod** is a compact type of antenna, which is often used in portable radio receivers. Ferrite is a magnetic material. It makes the lines of force of the received electromagnetic field tend to bunch together through the rod. This increases the apparent strength of the signal and makes the receiver more sensitive.

One or more coils of fine wire are wound on the rod and the magnetic fields induce currents in these coils. The coils are connected to variable capacitors that, together with the coils, form tuned circuits. These tuned circuits make the receiver more selective

RF amplifier: This includes a resonant circuit (often a combination of a capacitor and an inductor) that makes the receiver amplify one particular frequency very strongly. In other words, the tuned receiver has very *narrow* bandwidth in the radio-frequency band.

The tuned circuit is selective and does not amplify the signals from transmitters with slightly lower or higher frequencies. Usually there are two or three stages of tuned amplification, so increasing selectivity and, at the same time increasing sensitivity.

Demodulator: In an AM system, there are two stages in demodulation. The first stage is to rectify (p. 58) the output from the rf amplifier. The way this is done is shown in the drawing at the top of the opposite page. The effect of demodulating an audio transmission is shown in the drawing opposite.

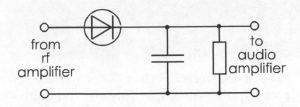

The second stage is to 'average out' or smooth the amplitude of the rectified signal. The capacitor does this. The result is an audio signal or data signal across the resistor. This is then sent to an amplifier.

Another way of thinking of this is to say that the capacitor acts as a low-pass filter, passing the low-frequency audio or data signal, but conducting the high-frequency carrier signal to the 0 V line.

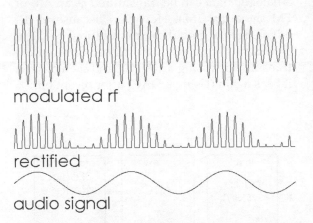

modulated rf

rectified

audio signal

Demodulation is different in an FM receiver. The receiver includes an rf oscillator that is tuned to the same frequency as the transmitter. A comparator circuit compares the frequency of the built-in oscillator with the frequency of the received radio signal. It produces a voltage that is proportional to the difference between the frequencies. The voltage rises and falls as the frequency of the received signal increases or decreases relative to the constant frequency of the oscillator. This voltage signal is a replica of the original audio signal.

Audio amplifier: The output from the demodulator may need amplification. A wide band amplifier is needed for an audio signal.

Output: A speaker is required for speech and music. A data transmission signal may first go to a decoder. After that it may go to various kinds of control circuit, to a computer screen, or to a memory for storage.

Radio astronomy

Events in distant galaxies generate electromagnetic waves at radio frequencies.

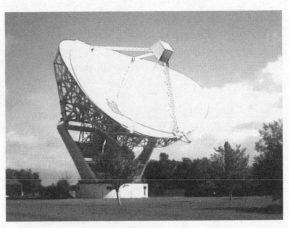

By receiving and analysing these signals, we can learn much about the Universe. Above is one of the radio telescopes at Jodrell Bank, Cheshire. The parabolic reflector concentrates the received radiation on to an antenna, which is mounted at its focal point. The rf amplifiers are also at this point. Signals from these are passed to computers for analysis.

Questions on radio reception

1 Draw a system diagram of an FM radio receiver and explain what each stage does.

2 Explain how an AM signal is demodulated.

3 At what stages in a radio receiver is there a (a) wideband amplifier, and (b) a narrow-band amplifier?

4 What features of a radio receiver make it more selective? Why is selectivity important?

41 Digital communications

Serial transmission

In computers and other data processing devices, data is usually handled in bytes. Eight parallel lines carry one bit each. Setting up eight separate transmission channels between a transmitter and a distant receiver is not practicable.

Normally, data is transmitted one bit at a time, or *serially*. The diagram shows exchange of data between two computers. A special logic interface IC receives bytes of data from the computer.

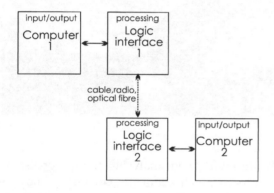

Usually the IC adds additional bits to this to prepare it for transmission. These include a start bit, a parity bit (see 'Parity' opposite) and a stop bit. The receiver is waiting (or 'marking') while the received voltage is low. The arrival of the start bit tells it that there are 9 more bits to follow, equal in length to the start bit. Reception begins when the start pulse arrives. From then on, the system clock in the receiver has to time only 9 bits.

Even if the receiver clock does not run at precisely the same speed as the transmitter clock, it is unlikely to get out of step during such a short transmission. Because the clocks do not need to be synchronised, this is known as an **asynchronous system**.

Analogue data

Analogue data includes speech, music, and video. It may also include data from sensors, such as when a satellite sends temperature readings back to Earth.

Analogue data can be transmitted as an AM or FM signal (p. 167). However, because of the advantages of digital communication (p. 172) it is better to convert it to digital form before transmission (using an ADC) and then convert it back again (Using a DAC) after reception.

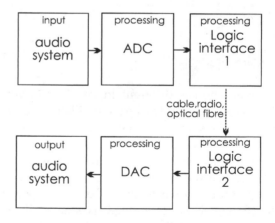

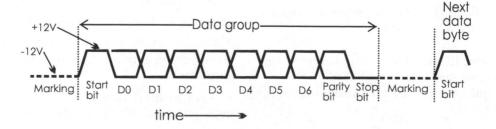

The first stage in analogue to digital conversion (ADC) is sampling. The signal is sampled at regular intervals and each sample is converted into its digital equivalent.

The drawing below shows the signal being converted into 4-bit samples but this gives only 15 possible values. For greater precision, most ADC converter ICs produce at least a 12-bit sample, which gives 4096 possible values. If the sampling is to keep track of the varying signal voltage, it is essential to take samples at short intervals. It can be shown that the sampling frequency should be at least double the highest frequency present in the signal. In the example in the diagram, the signal is sampled at 44 kHz. This just covers the top end of audio frequencies, at 20 kHz.

Parity

The data byte comprises seven data bits, plus one bit that it known as the **parity bit**. The parity bit is either '0' or '1', and is selected automatically so as to give an even number of '1's in the byte. For example, suppose the data bits are:

1101001

This has an even number of '1's so the parity bit is '0' and the complete byte is:

11010010

The byte is checked for parity by the logic interface at the receiver. If the byte arrives with an odd number of '1's, it shows that something is wrong.

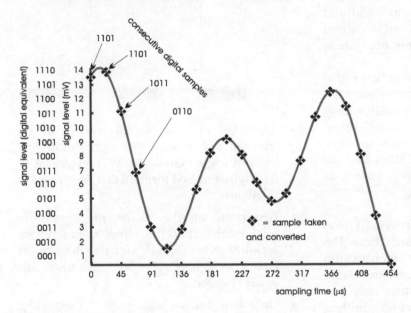

As another example, take these 7 data bits:

0110010

They have an odd number of '1's, so the parity bit must be '1' to make the number even. The completed byte is:

01100101

The parity described above is called **even parity**. Some systems use **odd parity**, in which the number of '1's is made odd. There are ICs that automatically check parity and indicate when there is an error.

The logic at the transmitter receives the output from the ADC. It then transmits it serially. At the other end of the channel, a similar IC in the receiver is able to collect the bits as they arrive and assemble them into bytes again.

It is important to check that there have been no errors in transmission. The only error that can occur is that a '0' may become a '1', or the other way about. The simplest way to check that this has not happened is to check the **parity** of the bytes as they arrive at the receiver.

Parity checking is a simple operation but is itself subject to error. For example, if *two* bits are wrong, the errors may cancel out. Other similar but more complicated techniques are used for error checking when accuracy is essential.

Self test

If parity is even, what parity bit is added to these seven bits:

0101100

Pulse code modulation

Pulses may be sent along a cable as a sequence of high and low voltage levels but, for transmission by radio, optical fibre and also by cable, they are more often modulated on to a carrier frequency. This is pulse code modulation, as described on p. 166. PCM is the most widely used technique for transmitting digital and (after conversion) analogue data. Bursts of the carrier frequency, representing '0's and '1's, are transmitted and then demodulated at the receiver.

The use of carriers means that many different messages may be sent along the same channel at the same time. This is a valuable feature. Communications links such as satellites and under-sea cables are expensive to build and maintain. The more signals that can be carried at any one time, the more economic the system.

The technique is to modulate the pulsed signal on to one of a number of carrier frequencies. The frequencies are spaced sufficiently widely apart that they do not interfere with each other. Yet they are close enough together to allow many carrier frequencies to be fitted into the bandwidth of the channel. This is known as **frequency division multiplexing**.

With FDM, the modulated carriers all pass through a single channel at the same time. The signals keep separate from each other because of their different carrier frequencies. Using only a single channel means that only one broadband amplifier is needed for transmitting and for receiving the multiplexed signals.

On reception, the individual tranmissions are separated out by using circuits tuned to each of the carrier frequencies. The separated signals are then routed to computers, and to the public telephone network. Digital signals that were originally audio or analogue signals are restored to their original form by analogue to digital converters.

Advantages of PCM

Pulse code modulation is the basis of today's 'information explosion'. Ever-increasing amounts of data are being transmitted across the world at ever increasing speeds. Below are some of the reasons for the success of PCM:

- Digital signals consist of '0's and '1's. There are no intermediate levels that may become difficult to recover after a signal has been distorted during transmission.

- Digital signals are highly immune from noise. The signal below, though noisy, is still recognisable as a pulsed signal.

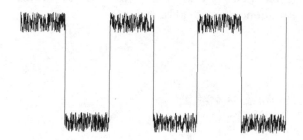

- Distorted or noisy signals can easily be 'cleaned up'. Passing the signal though a Schmitt trigger circuit (p. 103) restores it to its original pulsed form. In this way, noise is eliminated.

- Distortion of the shape of signals is unavoidable, but the effect is much less serious with digital signals than with analogue signals. The original pulses are easily restored.

- On a long transmission path (for example, England to Australia) the signal can be cleaned up, amplified and re-transmitted at a series of **regenerators** along the route. Digital signals are much easier to regenerate accurately than analogue signals. Fewer regenerators are required along the route.

- Digital signals are suited to processing by computer or digital logic circuits. This makes such procedures as automatic parity checking and multiplexing much easier. The channel can be completely under computer control.

Time division multiplexing

This is another technique for transmitting several signals along the same channel. Unlike frequency division multiplexing (p. 172), all transmissions are on the *same* frequency. Each can use the full bandwidth of the channel.

In TDM, signals from several sources are allocated to short time slots. Usually a time slot is about 1 s long. The signals from different sources are transmitted one after another in quick succession. Each signal is automatically placed in its allocated slot. At the receiving end the signals are sorted out automatically and routed to their correct receivers.

One of the complications of this system is that it has to be **synchronous**. The transmitter and receiver are synchonised so that time slots at the transmitting and receiving ends occur at exactly the right times. In contrast to this, frequency division multiplexing is **asynchronous**. The channels are independent and transmitters can send messages at any time.

TDM is used on the telephone network and in radio satellite communications. The equipment required for TDM is more complicated but the system is able to carry twice as much data as an equivalent FDM system.

RS-232 ports

The RS-232 standard defines a serial port for transmitting data over cables to a distance of up to 15 m. The system is often used for communications between computers. The standard specifies the type of connector as either a 9-pin or a 25-pin D-type connector. The logic levels are –12 V for digital '1' and +12 V for digital '0'. ICs are available for converting between TTL logic levels and RS-232 levels. The standard specifies that the connecting cable has nine lines. Three of these are concerned with sending the signal: SG, the signal ground line; TD, transmitted data; RD, received data. In this way, the system allows for two-way transmission.

Many simpler (not RS-232) systems use just these three lines.

In RS-232, the other six lines carry **handshaking** signals. These are high or low logic levels by which the two stations coordinate their actions. For example, a high level on the RTS line is a 'request to send' telling the receiver that the transmitter is waiting to send data. If it is ready to receive data, the receiver puts a signal on the CTS line, indicating 'clear to send'. On receiving this signal, the transmitter begins sending data on the TD line.

Questions on digital communications

1 What is PCM? What advantages does it have over analogue transmission?

2 What is frequency division multiplexing? What are its advantages?

3 Explain, giving an example of each, the difference between asynchronous and synchronous data transmission?

4 Outline the way in which an analogue signal, such as a musical signal, is converted to a stream of bits for transmitting by cable or radio.

5 If seven data bits are 0100100, and the parity is even, what will the eighth bit be?

6 Describe how a byte of data is transmitted asynchonously. What bits are added to the 8 data bits and what is their function?

7 What is time division multipexing?

42 Computers

Topics 32 to 34 describe the operation of microcontroller systems. Computers have much in common with microcontroller systems, but are designed for a different purpose. Microcontroller systems are relatively simple, and centre mainly on a single IC, the microcontroller. They are intended for controlling equipment and applicances such as dishwashers and mobile phones. Computers too are used in control systems, but mostly in highly complex systems. For example, computers are used to control power stations, jumbo jets, and radio telescopes. On the right is the main computer console that controls the electricity generators at Ironbridge Power Station, Shropshire.

For most people, a computer is the familiar personal computer or its more portable equivalent, the laptop computer. These are intended for use in offices, laboratories and the home. They are used for handling data, for calculating, for communicating on the Internet and for playing games, among many other applications. The diagram below shows the main parts of a PC system:

The microprocessor is the heart of the system. It corresponds to the main processing units of a microcontroller, including the arithmetic logic unit. A microprocessor has to communicate with all other parts of the system. It needs many address lines, data lines and control lines, which function as described on p. 145. The newer microprocessors, such as the Intel Pentium and the AMD K6, can address several gigabytes of memory and can handle 32-bit data. There are also many control lines. This means that the microprocessors need several hundred pins.

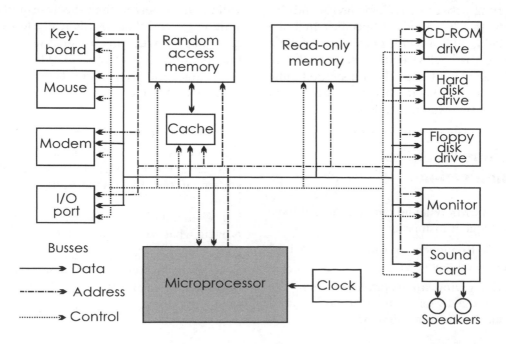

The action of the microprocessor is timed by the **system clock**. This is basically an astable, but one that runs at high speed. To process such large amounts of data at high speed, the clock in a typical PC runs at several hundred megahertz.

Memory

There are three types of memory:

RAM: Used for temporary storage of data and programs. Typically, there is 32 Mb or 64 Mb. Memory is often expandable by plugging in extra ICs. This might be necessary if you are working with extra-large files, such as detailed photographs.

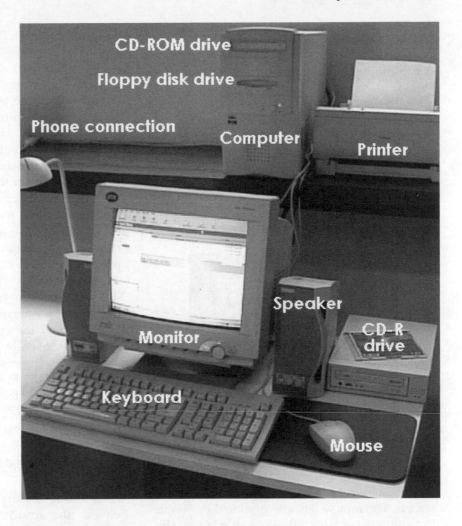

Memo

A **file** is a block of data or a program stored or transmitted in binary code. Files vary in length from one or two kilobytes to many megabytes.

ROM: For storing routines used at start-up.

Cache memory: This is a special type of RAM, which can be written to and read from very quickly. It is used for temporarily storing addresses and data that the processor might need at short notice. This helps to make processing faster. The cache memory may be on a separate chip or on the microprocesssor chip. Typically, there is 64 Kb of cache memory, but some computers have more.

Input and output

The computer needs to communicate with the outside world. It has many devices attached to it for this purpose. Some are for input and some are for output. Some are for both. Some are built into the main case of the computer while some are separate and connected to the computer by cable.

The input/output devices are placed on the left and right of the diagram opposite.

The photo above shows a typical PC system in the home. The main circuits and some of the input and output devices are housed in the tower at the rear. The phone connection can be seen running to a wall socket on the left.

Input devices

A typical PC system has a number of more-or-less standard input devices:

Keyboard: This is usually the main interface between the computer operator and the system. The microprocessor scans the keyboard frequently to detect key-presses. When keys are pressed, it responds accordingly.

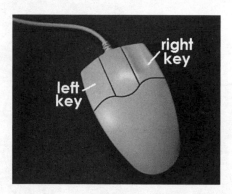

Mouse: A mouse (above) is the most convenient way of moving the image of a small arrow (known as the **mouse pointer**) over the screen. The mouse has one or more keys on it. In many programs, the operator uses the mouse to position the pointer over images of 'buttons' (below). Clicking one of the keys on a 'button' results in an appropriate response from the computer.

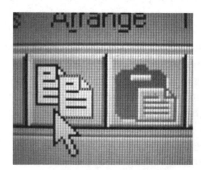

In the photo on p. 174, the operator is using a mouse (not a keyboard) to control the power station.

CD-ROM drive: The CD-ROM drive is a read-only drive and stores up to 6.5 Mb. The discs are the same size and use the same recording technique as musical compact discs (p. 162). Access to the CD-ROM is very fast, so software can be run direct from the drive.

CD-ROMs are much less destructible than floppy disks. In particular, they are unaffected by magnetic fields that so easily corrupt the data stored on a floppy disk. This makes them the present choice for the distribution of data and programs. The 'cover discs' of computer (and other) magazines are a popular application for CD-ROMs.

Most PCs have a build-in sound card that allows ordinary music CDs to be played on the drive, through the computer's sound system. Software is usually sold on CD-ROM. The major manufacturer's and retailers issue their catalogues on CD-ROM, often with the facility to link up with their web site for placing orders.

Output devices

These are the means by which the computer delivers the results of its activities. They include:

Monitor: Also known as a visual display unit (VDU). This is usually a cathode ray tube operating on digital signals from the computer to produce a coloured image. Laptops have a colour screen working on the principle of an LCD.

Sound card: Most people use this only as an output device. It can be used for listening to CDs played on the CD-ROM drive, for playing sound files downloaded from the Web or for listening to sounds that are part of a multimedia presentation. With suitable software, the sound card can be used to generate musical and other effects. The output goes to a pair of speakers. It can also be used for recording sound signals from a microphone, a tape, a disc or an MP3 player.

Input/Output devices

The remaining devices allow two-way communication.

Floppy disk drive: A floppy disk is a thin circular sheet of magnetic material in a plastic cover. Data is written or read using a magnetic head and circuits similar to those used in a tape recorder/player (p. 163). The floppy disk can be removed from the drive, so it is useful for backing up data, and for transferring data or programs to other computers. However, the rate of transfer of data is relatively slow. A floppy disk holds up to 1.44 Mb. This is not large enough for storing the larger computer files of present-day computers, so floppy disks are less useful than they once were.

CD-R drive: The CD-recordable drive allows data to be recorded on inexpensive CD-R blanks. Their capacity is 6.5 Mb, so a usefully large amount of data or software can be recorded on them. Some drives allow data to be recorded only once, but others allow the data to be overwritten with new data.

Modem: The name is short for **mo**dulator/**dem**odulator. It is a two-way device that connects the computer to the public telephone system. It is used for accessing the Internet, to view websites and to exchange e-mail. It is also used to send or receive files from other computers.

Ports: These are complex ICs that provide a means of connecting the computer to a range external devices. They may accept input or provide output or perhaps both at the same time.

Usually, a computer has at least one **parallel port**. The cable from a printer is usually connected here. Typically, this has a 36-pin connector and carries 8 bits at a time along 8 data lines. Parallel data transfer is fast enough to keep the printer fully occupied. Other lines are used for handshaking (p. 173) between the computer and printer.

There may also be one or two serial ports, which are often RS232 ports (p. 173). Being serial ports they transmit only 1 bit at a time and are slower than the parallel ports. Many add-on devices can be attached to these ports. Examples include a joystick for playing games, a microcontroller programmer (p. 149), and a digital camera.

Hard disk drive

This is used for fast storage of programs and data. The capacity of the drive is high, rated in gigabytes. It contains a stack of several disks turning on the same spindle. The disks are coated with a magnetic material. There is a pair of magnetic heads to read or write on the upper and lower sides of each disk. When the disks are spinning, the heads 'float' with a thin film of air between the disks and the heads. Being so close to the disk, the head is able to record a large amount data in a small area. The disk is not removable from the drive, being sealed in to exclude dust. This is essential, as the heads are so close to the disk that even a fine smoke particle between the head and the disk would cause the disk to crash.

The hard drive is the main storage unit of the computer. All programs currently in use are stored there, together with the files that they use or produce. When the computer is running, programs and blocks of data are continually being transferred between the hard disk and RAM.

Busses

The parts of the computer are connected together by sets of parallel lines (see p. 145). These sets are known as busses. The **data bus** has at least 8 lines so that data can be transferred a byte at a time. Often the bus has more lines, perhaps as many as 32. The **address bus** often has 24 or more lines so that many different addresses can be placed on the bus. The control bus has at least 6 lines.

43 Control systems

Electronic control systems are found almost everywhere, from controlling the ignition system of a car to the control of rush-hour traffic on the motorways. Electronics controls everything from the drilling machine on the factory floor to the coffee-making machine in the boardroom. Electronics controls everything from the fan-heater in the living-room to the power station that generates the electricity it uses.

There are so many electronic control systems that we can take only a few examples to illustrate the main types of control system that have been invented.

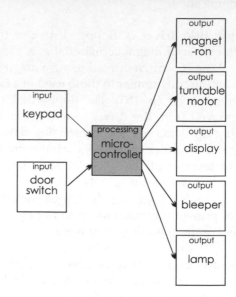

Microwave oven

This familiar domestic appliance has a fairly complicated system that makes it much easier to use than the ordinary gas or thermal electric oven.

The centre of the system is a microcontroller or possibly a microprocessor. This is responsible for running the system. The program for this is permanently stored in its ROM. In addition, it needs a small amount of RAM and possibly three clocks. One is the system clock, another is an elapsed time clock, which provides the cooking times, and the third gives the time of day. These clocks may be on separate ICs, though they might also be on the microcontroller chip.

The system has two inputs. The keypad is for entering cooking power and times and also has keys for starting, resetting and other functions. The door switch senses whether or not the door is closed.

There are five outputs. The magnetron generates the microwaves. This is switched fully on when cooking at high power, but is switched on for intermittent periods when cooking at reduced power or when defrosting. It can not be switched on when the door is open. The turntable motor runs for the whole cooking time. The display normally shows the time of day, but displays time settings as they are entered at the keypad, and the time remaining during cooking. It may also display power levels and messages. The lamp comes on when the door is open or when cooking is in progress. The bleeper sounds when keypad entries are made and at the completion of cooking.

The microcontroller unit accepts the inputs from the keyboard and door switch. It then goes through a logically controlled series of steps to cook the food as instructed. Its clocks help it to time the processing according to the instructions that have been keyed in.

This is a flowchart of a possible program for the microwave oven. Most of the steps are self-explanatory. Some of the steps omit details. For example, the 'Keypad entry routine' would be a long one, to allow for different types of entry such as cooking instructions, setting the clock, or using the clock as a kitchen timer. In the early stages of development it is best to keep the flowchart relatively simple so that its overall action can be more easily understood. Later, the programmer can come back and fill in the details. A routine for keypad entry could be used at several places in the flowchart. We call this a **subroutine**, a program within a program.

Note that this is a *digital* program. The door is open or closed, the lamp is on or off, the magnetron is on or off. Also there is plenty of scope for logic in the programming. A typical logical statement might be: 'IF the start button is pressed AND the door is shut, THEN start cooking now'. Microcontrollers are ideal for logical programs of this kind.

Start

Display time of day

Read keypad

Any key pressed? — No

↓ Yes

Keypad entry routine

Read keypad

Start button? — No

Yes

Read door switch

Door closed? — No

↓ Yes

Magnetron, lamp, & motor on

Start timing

Things to do

The flowchart above goes only as far as the beginning of cooking. Carry it on to when the bleeper sounds at the end.

Fan heater

Although the *hardware* of the control system of the fan heaters in the photo is mainly digital, the *action* involves analogue quantities. In fact, the heater could be controlled entirely by an analogue circuit. A few years ago, it would have been.

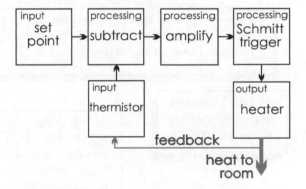

The system diagram (Note: not a flowchart) and the following description shows how the analogue version works.

input **set point** → processing **subtract** → processing **amplify** → processing **Schmitt trigger**

input **thermistor** → (into subtract)

processing Schmitt trigger → output **heater**

feedback (heater → thermistor)

heat to room

There are two inputs. One is from a sensor that registers the temperature of the air in the room. The other is from a control used by the operator to set the air temperature required. This is called the **set point**. The control might be a variable resistor wired as a voltage divider. The output voltage is proportional to the set point. The next stage finds which temperature is higher by subtracting actual temperature from the set point. For this we use an op amp wired as a comparator. Its output is amplified to increase sensitivity. If the result is positive, the Schmitt trigger switches on the heater.

The Schmitt trigger is used to avoid the heater being switched on and off in rapid succession when the air temperature is close to the set point. This would lead to excessive wear of the switch contacts.

Most of the heat is carried away to warm the air in the room. A small amount of heat finds its way to the thermistor. It is fed back to an earlier stage in the system. We call this action **feedback**. In this system, the action of the feedback is *negative*. If the air is too cool, it turns the heater on. If the air is too warm, it turns the heater off. This is **negative feedback**.

Negative feedback is often used in control systems. It acts to oppose change. It gives stability. It is often used in **regulator** systems like this one.

The heater in the photo is digital. It is controlled by a microcontroller. There is a keypad for entering the set point. Operations such as subtraction and the Schmitt trigger action are all controlled by logical programming. The microcontroller is also programmed for other actions that would not be possible with an analogue circuit.

Feedback in the heater system makes use of currents in the air outside the system. Similarly, if you are flying a radio-controlled model plane, you watch it as you manipulate the controls. In this case, feedback is visual. Most control systems have sensors to provide them with feedback. On p. 155 we showed how the automatic door system uses sensors to confirm that the doors have actually opened or closed.

Design time

Design programs by drawing them out as flowcharts. Here are some suggestions for topics taken from previous 'Design time' pages.

A microcontroller version of the twisty wire game, with a few novel features (pp. 121, 157).

A rain detector that sounds an alarm (p. 105).

An egg timer (p. 121) that is based on the internal timer of a microcontroller.

A sprinkler or pot-watering system based on a moisture sensor and solenoid operated valve. The system must not overwater the soil (pp. 129, 133).

A wind detector that records the number of gusts in 15 minutes (p. 105).

A model railway controller that does what is described on p. 133, and more. There could be three or four alternative routes, selected by keypad (pp. 133, 157).

A reaction tester (p.157)

A metronome circuit with LED and/or audio-'click' outputs (pp. 121, 143, 157).

Amplify your design by writing it as a program for a microcontroller. Try one of the simpler programs first. To complete the control system, build the input and output circuits it needs and program a microcontroller to run it.

A 'people-counter' (p. 143).

Voltage regulators

A circuit on p. 63 illustrates how to use a Zener diode as a voltage stabiliser. For more precise control of the voltage from a PSU we use a **voltage regulator**. This is an IC often contained in a 3-terminal package and looking like a power transistor.

The most popular 3-terminal regulators belong to the 78XX series. The last two digits of the number indicate the regulated voltage. For example, a 7805 regulator produces +5 V, and a 7812 regulator produces +12 V. The type number may also include a letter to indicate the maximum output current. For example, a 78L09 regulator is a low-power device producing 9 V at up to 100 mA. Other types are available that produce larger currents, up to 1 A is typical.

The connections for a 78XX regulator are:
- **Input:** This usually comes from a transformer, followed by a rectifier and smoothing capacitor, as on p. 59. The voltage must be 2.5 V to 6 V higher than the required regulated voltage. Special low dropout regulators operate with a supply voltage only 100 mV above the regulated output.
- **Common:** The 0 V line for both input and output.
- **Output:** Regulated output.

The standard circuit connections are shown below.

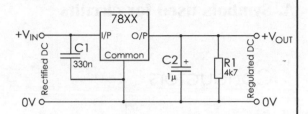

78XX regulators have an output precision of ±4%, which is better than that provided by a Zener regulator. The output voltage varies no more than 1% when the input voltage varies, provided it is 2.5 V or more higher. Similarly, the output voltage varies no more than 2% when variable amount of current are drawn from the regulator. The regulator almost completely removes any ripple from the supply.

The regulator has other important feature, such as **current limiting,** which drops the output voltage if excessive current is drawn from the regulator. There is also **thermal shutdown**, which cuts off the current if the regulator becomes overheated.

An equivalent series of regulators, the 79XX series, is used to regulate the negative supply to op amps and other devices.

Questions on computers and control systems.

1 Describe three input devices that are part of a typical computer system, and what they do.

2 In what ways does a microprocessor differ from a microcontroller?

3 Describe three types of memory found in a computer system, and explain the uses of each type.

4 Name three different devices (other than memory) used for storing data. What are their features and for what purposes are they used?

5 What is a bus in a computer system? What are the functions of the different kinds of bus?

6 Draw a flowchart of a computer program for controlling (a) a dishwasher, (b) a washing machine (only one wash program), (c) automatic sliding doors, (d) the entry barrier of an automatic car park, (e) a supermarket checkout, (f) disco lights or chaser lights, (g) a fan heater (as in p. 179, including an action like that of the Schmitt trigger).

7 Draw a flowchart program to replace the logic-gate system used to control the mixing tank on p. 133.

Supplements

A. Symbols used for circuits

Conductors

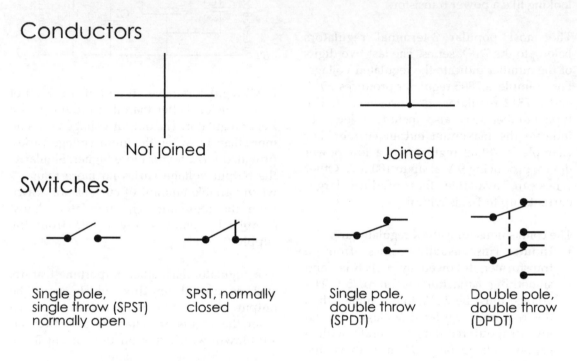

Not joined

Joined

Switches

Single pole,
single throw (SPST)
normally open

SPST, normally
closed

Single pole,
double throw
(SPDT)

Double pole,
double throw
(DPDT)

Push-buttons

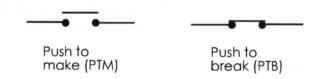

Push to
make (PTM)

Push to
break (PTB)

Power supply

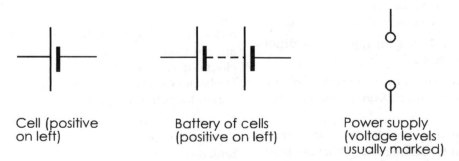

Cell (positive
on left)

Battery of cells
(positive on left)

Power supply
(voltage levels
usually marked)

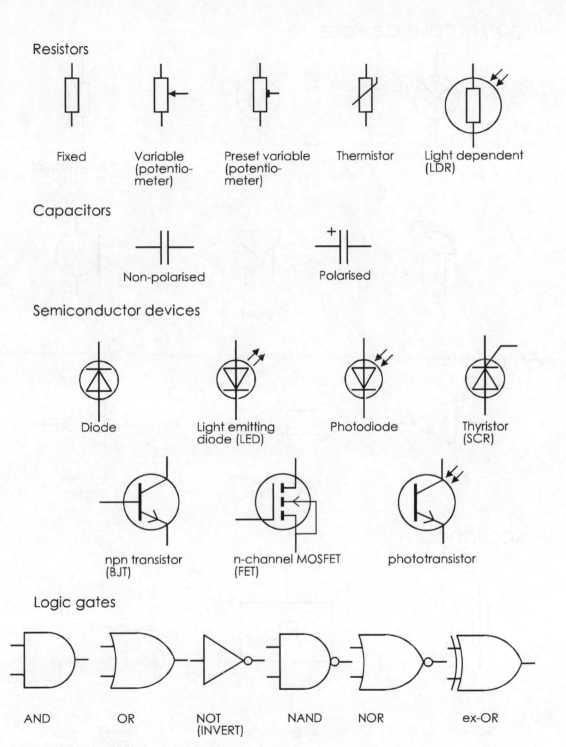

Resistors

Fixed

Variable
(potentio-
meter)

Preset variable
(potentio-
meter)

Thermistor

Light dependent
(LDR)

Capacitors

Non-polarised

Polarised

Semiconductor devices

Diode

Light emitting
diode (LED)

Photodiode

Thyristor
(SCR)

npn transistor
(BJT)

n-channel MOSFET
(FET)

phototransistor

Logic gates

AND

OR

NOT
(INVERT)

NAND

NOR

ex-OR

Input/output devices

Lamp

Motor

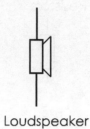

Loudspeaker

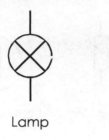

Bell

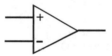

Buzzer

Microphone

Amplifiers

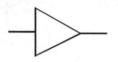

Amplifier

Operational
amplifier

Schmitt
trigger

Miscellaneous

Fuse

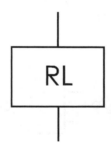

RL

Relay coil

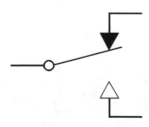

Relay contacts
(SPDT for example)

B. Symbols used in flowcharts

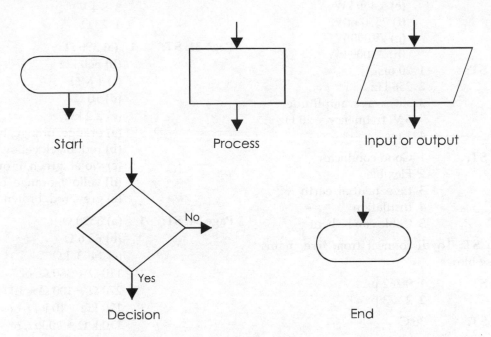

When using decision boxes, it is better for the 'No' arrow to point sideways, as shown above. The 'Yes' arrow continues the flow in the original direction.

C. Answers to questions

These are the answers to numerical Self Test questions and 'Questions on ...'. There are also answers to those questions that take only a few words.

Page 2: ST: They repel each other.

Page 3: ST: They become charged like the plastic, so are repelled.

ST: Because they would repel each other.

Page 6: ST: Cl^-, Na^+, SO_4^{--}.

Page 7: ST: Silver, copper, aluminium, gold, carbon, rubber.

Page 8: ST: By contact with air, by conduction to ground, by sparking to another object.

Page 11: ST:
    1 3
    2 4.5 V
    3 (a) Alkaline
       (b) Silver oxide
       (c) NiCad
       (d) Lead-acid

Page 13: ST:
    1 396 W
    2 43 mA
    3 180 mW
    4 500 mA
    5 (a) 34 000 mA
       (b) 1200 µA
       (c) 0.0012 A
       (d) 5.505 A
       (e) 0.058 mA

    6 (a) 4500 mV
       (b) 11 000 V
       (c) 0.675 V
       (d) 0.521 mV
       (e) 550 mV
       (f) 0.000 440 V
       (g) 0.022 mV
       (h) 3.300 kV

    7 (a) 0.675 kW
       (b) 33 000 kW

(c) 0.650 W
(d) 6 000 000 W
(e) 4.450 kW
(f) 2550 mW
(g) 79 000 W
(h) 33 000 kW

Page 17: ST:
1 20 ms
2 256 Hz
3 sine wave, amplitude = 6 V, frequency = 20 Hz
4 1 ms

Page 19: ST:
1 Good conductor
2 Flexible
3 Live, neutral, earth
4 Insulation
5 Cable can be bent

Page 21: ST: To disconnect from live mains when fuse blows

Page 23: ST:
1 0.462 p
2 22.73 p

Page 25: ST: 36 C

Page 26: ST:
1 2.5 W
2 1500 J

ST: –14 V, +14 V

Page 27: Questions about electricity
2 Polythene, acetate, glass, rubber
3 Positive and negative; negative
8 A flow of electrons (usually)
9 A battery is made up of two or more cells
12 Zinc-carbon or alkaline, 16 cells
Lead-acid, 12 cells
NiCad, 20 cells
13 Can produce a large current
14 (a) ampere, A, (b) volt, V, (c) watt, W
15 power = current × voltage
16 (a) μV, (b) MW
17 1.67 A
18 36 mW
23 1.32 p
25 Aluminium
33 4.242 V
35 96 C
36 2016 J
37 150 J

Page 29: ST:
1 50 Ω
2 885 Ω

ST:
1 2.5 Ω
2 7.5 mA
3 8.4 V
4 23 Ω

Page 31: ST:
1 (a) 390 Ω
(b) 560 kΩ
(c) 1 MΩ
(d) 10 Ω
(e) 2.2 kΩ
2 (a) orange, orange, black
(b) red, black yellow
(c) violet, green, brown
(d) yellow, orange, orange
(e) grey, red, brown

Page 34: ST:
1 (a) 3520 Ω
(b) 60.6 Ω
(c) 1403 kΩ
2 110 Ω + 360 Ω, or 270 Ω + 100 Ω + 100 Ω
3 120 kΩ + 10 k Ω, or 110 k Ω + 20 kΩ, or 100 k Ω + 30 k Ω

Page 36: ST:
(a) 20.34 Ω
(b) 1186 Ω
(c) 8.00 Ω
(d) 14.47 kΩ
(e) 50 Ω

ST:
1 (a) 39.3 Ω
(b) 928.5 Ω
2 (a) 1950 Ω
(b) 220 Ω

Page 37: ST:
1 (a) 33 kΩ
(b) 4.7 MΩ
(c) 2.2 Ω
(d) 1.8 Ω ± 5%
(e) 27 kΩ ± 10%
2 (a) 47R
(b) 100K
(c) 9K1
(d) 3K9J
(e) 750KK

ST: The multiturn trimmer is 5 kΩ ±10%

Page 39: Questions on the current and voltage rules
1 3.7A, away from the junction
2 0.2 A, away from the junction
3 R2

4 38.8 mA, 2.17 V, 4.65 V, 3.18 V

5 7.74 V

6 R1: 3 V, 30 mA. R2: 3 V, 63.8 mA

Page 39: Questions on voltage dividers.

1 4.78 V

2 11.95 V

3 2.45 V

4 6.28 V

5 4.66 V

6 0.78 V

7 13.34 V

8 R1 = R2

9 R1=150 Ω, R2 = 100 Ω, or in that proportion

Page 40: ST:　**1** 0.25 F

2 10 C

ST:　**(a)** 1 nF

(b) 2200 nF

(c) 1 000 000 000 nF

(d) 0.047 nF

(e) 56 000 nF

Page 43: SR:　**1** 556 μA, no current

2 Shorter

Page 46: ST:　**1** 240 V

2 300 turns

Page 48: ST:　**1** 3.6 W

2 63.7 mA

Page 49: ST:　**1** 65.5 W

2 26.45 Ω

Page 50: Questions on resistance and capacitance

2　**(a)** 560 Ω ± 2%

(b) 10 MΩ ± 20%

(c) 82 kΩ ± 5%

(d) 330 kΩ ± 10%

3　**(a)** brown, black, yellow, gold

(b) red, violet, black, no band

(c) brown, red, green, silver

4 4.3 kΩ, 4085 Ω to 4515 Ω

5　**(a)** 47 000 pF = 47 nF ± 5%

(b) 390 000 pF = 390 nF ± 10%

(c) 1000 pF = 1 nF ± 5%

6 1936 Ω

7 1820 nF or 1.82 μF

9 800 Ω

11 92 mA, away from the junction

12 15.7 mA, 1.885 V, 2.826 V, 1.287 V

13 2282 Ω

16 1.92 V

17 1600 turns, 100 Hz

18 B, commutator

21 2 Hz

22 4.64 V

23 1125 Ω

24 3.89 Ω

Page 53: ST: (a) Microswitch

(b) Toggle switch

(c) Tilt switch

(d) Microswitch

Page 60: ST: (a) 400 Ω (or nearest, 390 Ω)

(b) 1.3 kΩ

Page 63: Questions on diodes

2 Reverse bias, no current flows

Page 71: ST: 1 0.7 V

2 Close to 0 V

3 120

Page 81: ST: 1 250

2 1.25 S

Page 87: Questions on sensors and transistors

6 Collection, base, emitter, flows out of the emitter

8 Emitter current

12 470 Ω gives 21 mA

15 120

16 100, 0.7 V

23 4.5 mA

Page 102: ST:　**1** 4.2 V

2 Rises to 5.8 V

Page 108: ST: 1 373

2 (a) R1=10 kΩ, R2=2.2 MΩ

(b) R1=10 kΩ, R2=120 kΩ

(c) R1=100 kΩ, R2=300 kΩ

(d) R1=1 kΩ, R2=1.2 MΩ

Page 111: ST: 66 mV

Page 113: Questions on systems, sensors and interfacing

7 26 400

Page 120: Questions on timing

6 (a) 5.17 ms

(b) 2.42 μs

(c) 1137.4 s

7 (a) 27 kΩ

(b) 120 Ω

(c) 510 kΩ

8 (a) 8.056 ms
   (b) 4.761 ms
   (c) 3.243 ms

10 90%

Page 126: ST:

(a)

True gate

Input A	Output Z
0	0
1	1

(b)

NAND gate.

Inputs		Outputs	
B	A	Z1	Z2
0	0	0	1
0	1	0	1
1	0	0	1
1	1	1	0

(c)

TRUE gate.

Input A	Outputs	
	Z1	Z2
0	1	0
1	0	1

ST: (a)
.

No equivalent

Inputs			Outputs	
C	B	A	Z1	Z2
0	0	0	1	0
0	0	1	0	1
0	1	0	0	1
0	1	1	0	1
1	0	0	1	0
1	0	1	0	0
1	1	0	0	0
1	1	1	0	0

(b)

True gate

Input A	Outputs		
	Z1	Z2	Z3
0	1	1	0
1	0	0	1

Page 132: Questions on logic

3(a)

NOR gate

Inputs		Outputs		
B	A	B	A	Z
0	0	1	1	1
0	1	1	0	0
1	0	0	1	0
1	1	0	0	0

3 (b)

NOR gate.

Inputs		Outputs	
B	A	A+B	Z
0	0	0	1
0	1	1	0
1	0	1	0
1	1	1	0

10

Inputs		Outputs			
B	A	A+B	A.B	$\overline{A+B}$	Z
0	0	0	0	1	1
0	1	1	0	0	0
1	0	1	0	0	0
1	1	1	1	0	1

Exclusive-NOR

12

Inputs B A		Outputs				
		$\overline{B}$	$\overline{A}$	$\overline{A.B}$	$\overline{A}.\overline{B}$	Z
0	0	1	1	0	0	1
0	1	1	0	1	0	0
1	0	0	1	0	1	0
1	1	0	0	0	0	1

Exclusive-NOR

14 $Z = \overline{A} + B$

15

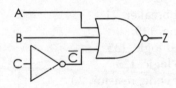

Page 140: ST: (a) 1000 0100
(b) 0011 0011
(c) 0010 1001
(d) 0011 0110 0010

Page 142: Questions on logical sequences
10 (a) 100 0101
(b) 0110 1001

Page 147: Questions on storing data
1 binary digit, 0 or 1
2 16
3 5120
5 The temperature is not greater than 25°C. Note that this is *not* the same as 'The temperature is less than 25°C'.
6 256
7 Writing value '9' into address 6

Page 156: Questions on programming
9 (left) Process, (right) input or output

Page 171: ST: 1

Page 173: Questions on digital communications.
5 0

Acknowledgements

The author thanks these companies and organizations for permission to take photographs on their sites:

Bossong Engineering Pty. Ltd., Canning Vale, Western Australia (p. 1).
Eastern Generation Ltd., Ironbridge, Shropshire (p. 18, p. 174).
Nuffield Radio Astronomy Laboratories, Jodrell Bank, Cheshire (p. 1, p. 169).

All photographs and all drawings (except those on pp. 2, 40, 44-45 and 163) are by the author.

INDEX

Index by type numbers

(See also the BJT data on p. 85.)